SpringerBriefs in Molecular Science

Biometals

Series Editor

Larry L. Barton

For further volumes:
http://www.springer.com/series/10046

Antonella Furini
Editor

Plants and Heavy Metals

 Springer

Antonella Furini
Department of Biotechnology
University of Verona
Ca Vignal 1, Strada Le Grazie 15
37134 Verona
Italy

ISSN 2191-5407   ISSN 2191-5415 (electronic)
ISBN 978-94-007-4440-0   ISBN 978-94-007-4441-7 (eBook)
DOI 10.1007/978-94-007-4441-7
Springer Dordrecht Heidelberg New York London

Library of Congress Control Number: 2012936759

Printed on acid-free paper

Springer is part of Springer Science+Business Media (www.springer.com)

Preface

Heavy metals are chemical elements with a specific gravity greater than 5.0. Among the 90 naturally occurring elements, 21 are non-metals, 16 are light metals, and the remaining 53 are (with As included) heavy metals. The definition thus includes the majority of naturally occurring elements, which from a biological perspective is not very helpful. However, only a limited number of heavy metals are soluble under physiological conditions and thus bioavailable to living organisms. Some of these are considered essential for life, including Fe, Co, Mo, Mn, Zn, V, Ni, Cr, Cu, and W. They are required as micronutrients or trace elements because they often act as cofactors in biochemical reactions, but they are toxic when present in excess. Other heavy metals such as Cd, Hg, Ag, Pb, and U have no known biological function and are toxic even at very low concentrations.

Heavy metals are often present as natural components of ultramafic or calamine soils, but the prevalence of heavy metals in the environment has increased more recently as a result of human activity. Metal processing facilities, mines, refuse dumps, sewage sludge, and traffic are all sources of heavy metals. In addition, the intense use of phosphate fertilizers and municipal sewage sludge in agriculture contributes to the accumulation of heavy metals in soils. The increasing concentrations are potentially toxic to both animal and plant life.

Metal mobility in the soil is strongly influenced by root exudates and microbes in the rhizosphere. Mobilized metals bind to root cell walls and are then taken across the plasma membrane by transport systems. Diverse families of metal transporters are induced under metal deficiency conditions, indicating their involvement in the regulation of metal uptake, transport, and distribution in the aerial parts of the plant.

For most plants, both essential and non-essential heavy metals cause toxicity symptoms and growth inhibition when present in excess. Heavy metals induce oxidative stress by generating free radicals and reactive oxygen species; they displace essential ions from proteins and other molecules; and then bind strongly to oxygen, nitrogen and sulfur groups, and hence inactivate enzymes by binding to cysteine residues.

Plants use various strategies to prevent heavy metals accumulating at sensitive sites within the cell, thus avoiding the damaging effects of heavy metal toxicity. One important detoxification strategy is the chelation of metals by a ligand, which is subsequently compartmentalized. Different metal-binding ligands have been recognized in plants and they appear to regulate different stages of metal transport and storage. These include organic acids, amino acids, peptides, and proteins.

Recent advances in molecular biology, sequencing technology, and bioinformatics mainly focusing on model plants, have rapidly increased our understanding of heavy metal detoxification and metal stress tolerance in plants. Understanding the molecular basis of heavy metal tolerance in plants will play a more prevalent role in food safety, and will also provide new strategies to address micronutrient deficiencies through the development of biofortified food crops accumulating higher levels of essential heavy metals such as zinc. Furthermore, to better understand the unique feature of several plant taxa of accumulating exceptional concentrations of heavy metals in aerial tissues will consent to improve the phytoremediation of contaminated sites. The following chapters report a broad overview of plant mechanisms involved in the transport, accumulation, and detoxification of heavy metals and highlight future prospects for the exploitation of heavy metal tolerance in plants.

Verona, April 2012   Antonella Furini

Contents

Contributors

Giovanni DalCorso Department of Biotechnology, University of Verona, Ca Vignal 1, Strada Le Grazie 15, 37134 Verona, Italy, e-mail: giovanni.dalcorso@ univr.it

Elisa Fasani Department of Biotechnology, University of Verona, Strada Le Grazie 15, 37134 Verona, Italy, e-mail: elisa.fasani@univr.it

Antonella Furini Department of Biotechnology, University of Verona, Strada Le Grazie 15, 37134 Verona, Italy, e-mail: antonella.furini@univr.it

Anna Manara Department of Biotechnology, University of Verona, Strada Le Grazie 15, 37134 Verona, Italy, e-mail: anna.manara@univr.it

Andrea Nesler Department of Biotechnology, University of Verona, Strada Le Grazie 15, 37134 Verona, Italy, e-mail: andrea.nesler@univr.it

Abbreviations

ABC	ATP binding cassette
ACC	1-aminocyclopropane-1-carboxylate
ADP	Adenosine diphosphate
APS	ATP sulfurylase
APX	Ascorbate peroxidase
AsA	Ascorbic acid
ATP	Adenosine triphosphate
ATP-PRT	ATP phosphoribosyl transferase
CaCA	Ca^{2+}/cation antiporter
CAT	Catalase
CAX	Cation exchanger
CCX	Calcium cation exchanger
CDF	Cation diffusion facilitator
CTR	Copper transporter
DHA	Dehydroascrobate
DHAR	DHA reductase
γ-ECS	γ-glutamylcysteine synthetase
GDH	Glutamate dehydrogenase
CgS	Cystathionine gamma synthase
GPX	Glutathione peroxidase
GS	Glutathione
GSH	Reduced glutathione
GSR	Glutathione reductase
GSS	Glutathione synthetase
GSSG	Oxidized glutathione
GST	Glutathione-S-transferase
HMA	Heavy metal-transporting P-type ATPases
HMW	High-molecular weight
HSP	Heat shock protein
IAA	Indole acetic acid
IRT	Iron-regulated transporter
JA	Jasmonic acid

kDa	Kilo Dalton, measure unit for protein weight
LMW	Low molecular weight
HMW	High molecular weight
LTP	Lipid transfer protein
MAPK	Mitogen-activated protein kinase
MATE	Multidrug and toxic compound extrusion
MDHA	Monodehydroascorbate
MDHAR	MDHA reductase
MHX	Mg^{2+}/H^+ exchanger
MRP	Multi-drug resistance-associated proteins
MT	Metallothionein
MTP	Metal tolerance protein
NA	Nicotianamine
NAS	Nicotianamine synthase
NRAMP	Natural resistance-associated macrophage protein
OAS	O-acetyl-L-serine
OEC	Oxygen evolving complex
OPT	Oligopeptide transporters
PAL	Phenylalanine ammonia lyase
PC	Phytochelatin
PCS	PC synthase (γ-glutamylcysteine dipepticyl transpeptidase)
PDR	Pleiotropic drug resistance
pH	Negative logarithmic of the hydrogen ion (H^+) concentration
PRX	Peroxidases
PSI	Photosystem I
PSII	Photosystem II
ROS	Reactive oxygen species
RuBisCO	Ribulose-1,5-bisphosphate carboxylase oxygenase
SA	Salicylic acid
SAP	Stress-associated protein
SMT	Selenocysteine methyltransferase
SOD	Superoxide dismutase
YSL	Yellow-stripe 1-like
ZAT	Zinc transporter of *Arabidopsis thaliana*
ZIP	ZRT-, IRT-related proteins
ZRT	Zinc regulated transporter

Chapter 1
Heavy Metal Toxicity in Plants

Giovanni DalCorso

Abstract Plants are sessile organisms that must cope with the surrounding soil composition in order to survive and reproduce. Soils often contain excessive levels of essential and non-essential elements, which may be toxic at high concentrations depending on the plant species and the soil characteristics. Many metals share common toxicity mechanisms, and plants deal with these metals using similar scavenging pathways. The impact of metal toxicity is made more complex by competition, since high levels of one metal may imbalance the uptake and transport of others, therefore contributing to the toxicity symptoms. Here, the toxicity symptoms and mechanisms of the most common essential and non-essential heavy metals will be considered.

Keywords Heavy metal · Plant nutrition · Metal pollution · Metal toxicity

1.1 Heavy Metals: Nutrients or Toxic Elements?

Plants acquire mineral elements from soil primarily in the form of inorganic ions. The extended root apparatus and its ability to absorb ionic compounds even at low concentrations makes mineral absorption highly efficient.

Mineral elements can be divided into two groups: essential nutrients and toxic non-nutrient elements. The essential minerals include the *macronutrients* nitrogen (N), potassium (K), calcium (Ca), magnesium (Mg), phosphorous (P), sulfur (S) and

G. DalCorso (✉)
Department of Biotechnology, University of Verona, Ca Vignal 1,
Strada Le Grazie 15, 37134 Verona, Italy
e-mail: giovanni.dalcorso@univr.it

A. Furini (ed.), *Plants and Heavy Metals*, SpringerBriefs in Biometals,
DOI: 10.1007/978-94-007-4441-7_1, © DalCorso 2012

silicon (Si), and the *micronutrients* chlorine (Cl), iron (Fe), boron (B), manganese (Mn), sodium (Na), zinc (Zn), copper (Cu), nickel (Ni), and molybdenum (Mo). These are essential components of plant metabolism and structure, and their absence or deficiency reduces fitness and inhibits growth and reproduction. Micronutrients are required in only small quantities and their excessive abundance in the soil (especially Cu, Ni and Zn), due to natural occurrence or introduction by anthropogenic activities, is also detrimental to the majority of plant species. Other minerals such as cadmium (Cd), mercury (Hg), lead (Pb), chromium (Cr), arsenic (As), silver (Ag), and antimony (Sb) are toxic to plants even at low concentration. These metals, collectively defined as *heavy metals* since their density is higher than 5.0 g cm^{-3}, are not considered to be nutrients because they have no known function in plant metabolism and appear to be more or less toxic to both eukaryotic and prokaryotic organisms (Sanità di Toppi and Gabbrielli 1999). Only recently, a carbonic anhydrase has been shown to bind Cd as a cofactor in the marine diatom *Thalassiosira weissflogii* (Lane and Morel 2000).

When studying heavy metal toxicity in plants, researchers must take into account the nature of the pollution phenomenon. First, the stress caused by contaminated soils is permanent, and therefore long-term rather than short-term molecular responses must be considered. Most studies have been carried out in hydroponic or in vitro culture, and have involved the application of extremely high metal concentrations in the growth media. This seldom resembles actual environment and represents the consequences of acute stress caused by a single metal species. Second, the toxicity of a heavy metal depends on its oxidation state, e.g., Cr(VI) is considered the most toxic form of Cr, and usually occurs associated with oxygen as chromate (CrO_4^{2-}) or dichromate ($Cr_2O_7^{2-}$) oxyanions. Cr(III) is less mobile, less toxic, and predominantly bound to organic matter in soil and aquatic environments (Shanker et al. 2005). Third, the ability of heavy metals to persist in the soil in the form that is bioavailable to roots (i.e., soluble and ready for absorption) is influenced by their adsorption, desorption, and complexation in the soil matrix, processes that are strongly influenced by soil pH, composition, and structure. Heavy metals tend to be more mobile in acidic soils. Finally, heavy metal toxicity is species dependent. For instance, metal-tolerant plants and certain plants known as hyperaccumulators [able to accumulate at least 100 mg g^{-1} (0.01% dry weight) Cd, As, and some other trace metals, 1,000 mg g^{-1} (0.1% dry weight) Co, Cu, Cr, Ni, and Pb and 10,000 mg g^{-1} (1% dry weight) Mn and Ni (Reeves and Baker 2000; Watanabe 1997) have defense mechanisms that avoid damage caused by heavy metal-induced stress, although the duration and magnitude of exposure and other environmental conditions, contribute to heavy metal sensitivity (Sanità di Toppi and Gabbrielli 1999).

Metal toxicity is also greatly influenced by the coexistence of many metals in the soil, which could have both *synergic* and *antagonistic* effects depending on the relative concentrations and other soil properties (i.e., presence of nutrient elements). For example, Ca^{2+} strongly inhibits the uptake of Ni in *Arabidopsis bertolonii*, whereas the opposite effect is seen in *Berkheya coddii* (Gabbrielli and Pandolfini 1984). Ni can induce Fe deficiency either by retarding its uptake or trapping Fe in the roots (Mysliwa-Kurdziel et al. 2004), and these effects can be partially overcome by supplementation with Mg (or Fe) ions suggesting a competitive interaction (Le Bot et al. 1990). In *Pisum sativum*, Mn toxicity can be reduced by applying indole acetic acid (IAA). This also promotes seedling growth, and it has been suggested that IAA protection is mediated by regulation of both the ammonium content and the activities of enzymes involved in ammonium assimilation (Gangwar et al. 2011). Mn toxicity is also reduced in the presence of Si. The biochemical and physiological basis of this phenomenon is poorly understood and may involve the modification of metabolic stress responses (Führs et al. 2009) and a change in the apoplastic Mn-binding properties that lead to a reduction in the concentration of Mn in the apoplast (Horst et al. 1999). As already stated, different species respond to combinations of ions in different ways. As an example, pea plants are protected from Cd toxicity by Ca, which limits Cd accumulation, whereas in *Brassica juncea*, Ca promotes the accumulation of As but reduces its toxicity (Rai et al. 2011a).

Fertilization is also known to influence heavy metal toxicity. The addition of phosphate reduces As toxicity in field-grown *Medicago truncatula* and *Hordeum vulgare* without modifying the specific uptake of As(V), and this may be due to the higher phosphate concentration into cells that outcompetes As in metabolic reactions (Christophersen et al. 2009). The accumulation and translocation of As in rice plants is inhibited when sulfur is abundant but enhanced when its availability is limited. This may reflect the prominent role of sulfur, which is a component of PCs and GS (both of which form complexes with heavy metals) and the impact of its availability on the synthesis of thiolic compounds, elements that ultimately affect As accumulation and metabolism (Zhang et al. 2011).

1.2 Toxicity Mechanisms of Heavy Metals

Heavy metal toxicity in plants occurs through four major mechanisms:

(1) Induction of oxidative stress and changes in the cell membrane permeability and integrity. Many heavy metals induce the formation of ROS such as H_2O_2, $O_2^{\cdot-}$ and OH$^{\cdot}$, which may be a direct process (via the Fenton and Haber–Weiss reactions, as shown in Fig. 1.1) or as secondary effect due to their toxicity into the cell. ROS have a negative impact on plant cells, for instance by inhibiting water channel and transporter proteins and enhancing lipid peroxidation. The latter alters membrane fluidity, stability, and structure, inhibiting membrane-dependent processes such as electron flow in chloroplasts and mitochondria. ROS are counteracted by the activation of *antioxidant enzymes* such as SOD, APX, GPX, CAT, and GSR, whose reaction mechanisms are shown in Fig. 1.2.

(2) Reaction with sulfhydryl groups (–SH). Heavy metals have a strong affinity for –SH groups [Cd, for example, shows a threefold higher affinity for –SH groups than Cu (Schützendübel and Polle 2002)] and therefore bind to structural

Fig. 1.1 Generation of ROS by heavy metals, including examples of reactions catalyzed by Fe (Halliwell and Gutteridge 1984)

Haber-Weiss cycle

$$H_2O_2 + OH^{\cdot} \longrightarrow H_2O + O_2^{\cdot-} + H^+$$
$$H_2O_2 + O_2^{\cdot-} \longrightarrow O_2 + OH^{\cdot} + OH^-$$

Fenton Reaction

$$H_2O_2 + Fe^{2+} \longrightarrow Fe^{3+} + OH^{\cdot} + OH^-$$
$$H_2O_2 + Cu^+ \longrightarrow Cu^{2+} + OH^{\cdot} + OH^-$$

Example of Fe catalyzed ROS production

$$O_2^{\cdot-} + Fe^{3+} \longrightarrow Fe^{2+} + O_2$$
$$Fe^{2+} + H_2O_2 \longrightarrow Fe^{3+} + OH^{\cdot} + OH^-$$
$$Fe^{3+} + H_2O_2 \longrightarrow Fe^{2+} + O_2^{\cdot-} + H^+$$
$$Fe^{2+} + OH^{\cdot} \longrightarrow Fe^{3+} + OH^-$$

proteins and enzymes containing them. This can prevent correct folding, interfere with catalytic activity, and perturb enzyme-mediated redox regulation (Hall 2002).

(3) Similarity to biochemical functional groups. As(V) in arsenate (AsO_4^{3-}), for example, is an analog of the micronutrient phosphate (PO_4^{3-}) and competes with it in many cellular functions. AsO_4^{3-} displaces phosphate in ATP, leading to the formation of the unstable complex ADP-As that interferes with the energy flows in the cell (Meharg and Hartley-Whitaker 2002).

(4) Displacement of essential (cat)ionic cofactors in enzymes and signaling components. Metal ions in the active sites of enzymes can be displaced by heavy metal ions resulting in the loss of activity, e.g., the displacement of Cu, Zn, Fe, and Mn cofactors from superoxide dismutase by Cd. The displacement of the ionic cofactors from signaling proteins (e.g., calmodulin and transcription factors) results in aberrant proteins that may perturb gene expression. This process can also interfere with homeostatic pathways for essential metal ions (Roth et al. 2006). For example, the displacement of Cu and Fe from proteins releases free ions that may cause oxidative damage, e.g., via Fe/Cu-catalyzed Fenton reactions (DalCorso et al. 2008).

The large number of targets for heavy metal toxicity means that negative effects tend to be firstly observed in those cells that are exposed first, i.e., cells responsible for the metal uptake. Heavy metals interfere with ionic homeostasis and enzyme activity, and these effects are apparent in physiological processes involving single organs (such as nutrient uptake by the roots) followed by more general processes such as germination, growth, photosynthesis, plant water balance, primary metabolism, and reproduction. Indeed, visible symptoms of heavy metal toxicity

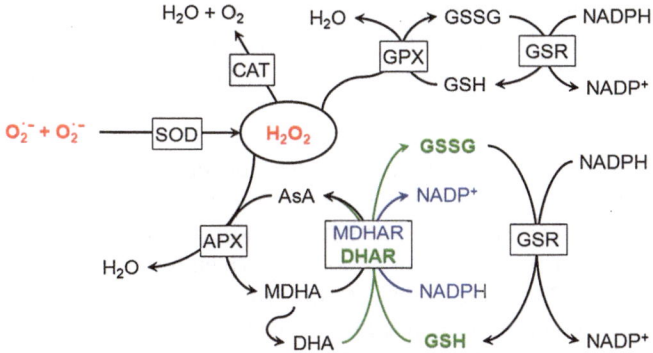

Fig. 1.2 Antioxidant enzymes responsible for the detoxification of H_2O_2 and $O_2{}^{\cdot-}$. *APX* ascorbate peroxidase, *AsA* ascorbate, *CAT* catalase, *DHA* dehydroascorbate, *DHAR* dehydro-ascorbate reductase, *GSR* glutathione reductase, *GSH* reduced glutathione, *GSSG* oxidized glutathione, *GPX* glutathione peroxidase, *MDHAR* monodehydroascorbate reductase, *NADP* nicotinamide adenine dinucleotide phosphate, reduced (NADPH) and oxidized (NADP⁺), *SOD* superoxide dismutase

include chlorosis, leaf rolling and necrosis, senescence, wilting and stunted growth, low biomass production, limited numbers of seeds, and eventually death. We now consider the effects of the most relevant heavy metals individually.

1.3 Non-Essential Heavy Metals: Cadmium, Mercury, Lead, Chromium and Arsenic

1.3.1 Cadmium

Cadmium (Cd) is one of the most phytotoxic heavy metals because it is highly soluble in water and promptly taken up by plants. This also represents its main entry into the food chain, making it a threat to human health. Even at low concentrations, the uptake by roots and the transport of Cd to vegetative and reproductive organs have a negative effect on mineral nutrition, homeostasis, growth and development (DalCorso et al. 2010).

In root cells, Cd imbalances water and nutrient uptake, interfering with the absorption of Ca, Mg, K, and P. It inhibits root enzymes involved in nutrient metabolism, such as Fe(III) reductase, nitrate reductase, nitrite reductase, glutamine synthetase, and glutamate synthetase, leading to Fe(II) deficiency, and reduced nitrogen assimilation and metabolism (glutamine and glutamate synthetases are responsible for the incorporation of ammonium into the carbon skeleton, DalCorso et al. 2008). Nitrogen fixation and primary ammonia assimilation is also inhibited in the nodules of soybean plants in the presence of Cd (Balestrasse et al. 2003). Cd inhibits root growth and lateral root formation, with the concomitant differentiation of numerous root hairs, for instance, in *Arabidopsis* and tobacco (Farinati et al. 2010). Tomato roots exposed to Cd are thicker and stronger (Chaffei et al. 2004). In shoot tissues, the most evident symptoms of Cd toxicity are leaf roll, chlorosis, water uptake imbalance, and stomatal closure (Clemens 2006). Chlorosis may reflect changes in the Fe:Zn ratio that negatively affect chlorophyll metabolism (Chaffei et al. 2004). Cd causes stomata to close independent of water status probably because its similarity to Ca allows Cd to enter guard cells through voltage-dependent Ca^{2+} channels and to mimic Ca^{2+} activity in the cytosol (Perfus-Barbeoch et al. 2002). Indeed, stomatal closure can be actively driven by Ca^{2+} accumulating in the guard cell cytosol. The increase in cytosolic free Ca^{2+} causes plasma membrane anion and K_{out}^{+} channels to open, resulting in the loss of water and turgor that drives stomatal pore closure (MacRobbie and Kurup 2007).

Both cellular and organellar metabolism are compromised by Cd. In chloroplasts, Cd damages the photosynthetic apparatus, targeting the light-harvesting complex II and the two photosystems (PSI and PSII) which are particularly sensitive. This reduces the chlorophyll and carotenoid content, increases non-photochemical quenching and limits both photosynthetic efficiency and effective quantum yield (Sanità di Toppi and Gabbrielli 1999). Moreover, by inhibiting enzymes involved in CO_2 fixation, Cd reduces carbon assimilation (Perfus-Barbeoch et al. 2002). Cd also affects sulfur metabolism in the chloroplasts by inducing the accumulation of thiolic compounds with a concomitant reduction in leaf ATP-sulfurylase and O-acetylserine sulfurylase activity, i.e., the first and the last enzymes in the sulfate assimilation pathway (Astolfi et al. 2004).

Cd is toxic at the cellular level by interfering with mitosis and inhibiting cell division, due to chromosomal aberrations and inhibition of mitotic processes (Benavides et al. 2005). In *Arabidopsis*, Cd induces mutations, leading to floral anomalies, embryonic malformations, and poor seed production (DalCorso et al. 2008).

Although Cd does not take part in the Fenton and Haber–Weiss reactions, without Cd ions altering their oxidation state (Clemens 2006), exposure can still induce oxidative injuries such as protein carbonylation and lipid peroxidation, disrupting cell homeostasis and interfering with membrane functions (Romero-Puertas et al. 2002; Schützendübel et al. 2001). This appears to reflect a Cd-induced imbalance in the activity of antioxidative enzymes, CAD and SOD *in primis*, leading to the accumulation of ROS, which may be a general effect of redox imbalance or a specific response to heavy metal stress (Romero-Puertas et al. 2004). Other plants induce GDH in response to Cd (Boussama et al. 1999). GDH activity is correlated with the onset of senescence and associated changes in nitrogen metabolism (Masclaux et al. 2000). Similar changes in nitrogen metabolism are observed in plants exposed to Cd so it is possible that the toxic effects of Cd reflect the induction of senescence. In peroxisomes, Cd induces glyoxylate cycle enzymes (malate synthase and isocitrate lyase) as well as peroxisomal peptidases, the latter being well known as leaf senescence-associated factors (Chaffei et al. 2004).

A secondary effect of ROS accumulation is the perturbation of signaling pathways mediated by H_2O_2 and oxygen radicals. Indeed, H_2O_2 plays a role as signal molecule in triggering, for instance, defence mechanisms against both abiotic stresses (Dat et al. 2000; Sharma et al. 1996) and pathogen attack (Bestwick et al. 1998; Thordal-Christensen et al. 1997). Interfering with H_2O_2 accumulation, Cd meddles with the signal transduction pathways in which ROS are involved. Cd^{2+} can also displace the chemically similar Zn^{2+} from zinc finger transcription factors, thus interfering with gene expression (Sanità di Toppi and Gabbrielli 1999). Similarly, Cd^{2+} can displace Ca^{2+} from calmodulin proteins, thus perturbing intracellular calcium level and altering calcium-dependent signaling, e.g., the regulation of stomatal closure discussed above (DalCorso et al. 2008).

1.3.2 Mercury

Mercury (Hg) is generally found only in trace concentrations in soil, and it is tightly bound to organic matter and clay particles or as a sulfide precipitates (Schuster 1991). The predominant source of Hg in the soil is from mining and industrial waste (Zhou et al. 2007). The toxicity of Hg depends on its chemical state (e.g., HgS, Hg^{2+}, Hg^0, and methyl-Hg). The predominant form in agricultural soils is Hg^{2+}, which is not particularly phytotoxic at normal concentrations, but it is soluble, highly reactive, and readily taken up by plants (Han et al. 2006). Alternatively, the uncharged and volatile form Hg^0 can enter leaves via the stomata and diffuse to the mesophyll cells where it is oxidized to Hg(II) (Zhou et al. 2007). The first visible symptoms of Hg toxicity are the profound inhibition of root and shoot growth (Cho and Park 2000). The molecular basis of Hg phytotoxicity remains uncertain but probably reflects: (i) the affinity of Hg for –SH groups; and (ii) the direct generation of ROS via the Fenton reaction, which in turn induces oxidative stress (Fig. 1.1).

Roots show the first signs of Hg toxicity because these are the first tissues to be exposed to the metal. The suppression of root growth by Hg has been observed in tomato seedlings, in *Brassica* spp. and in *Spinacia oleracea* (Cho and Park 2000; Ling et al. 2010). At high concentrations, Hg can bind to water channels in the plasma membrane, interfering with water flow and stomatal functions. When wheat root cells were exposed to $HgCl_2$, the hydraulic conductivity of the membranes was reduced, suggesting that membrane depolarization may inhibit water transport (Zhang and Tyerman 1999). Hg also strongly inhibits photosynthesis by interacting with metal ions in the PSII proteins D1 and D2 and with the Mn-cluster of the OEC (Patra et al. 2004). Oxygen evolution and thylakoid electron transport are also inhibited because Hg depletes the 33-kDa manganese stabilizing protein on the luminal side of PSII (Bernier and Carpentier 1995). PSI is also compromised by Hg, which oxidizes the P_{700} chlorophyll *a* when present as $HgCl_2$ (Sersen et al. 1998). In addition to Hg-induced chlorophyll depletion, these negative effects eventually result in a dramatic reduction in the photosynthetic quantum yield (Cho and Park 2000).

Laboratory experiments with various explants have shown that high concentrations of Hg are genotoxic, causing chromosomal damages, interfering with mitosis and meiosis, and inducing polyploidy (Patra et al. 2004).

Hg has a global impact on the redox state of the cell because it catalyzes the formation of ROS. In tomato seedlings, exposure to Hg induces the formation of H_2O_2 (Cho and Park 2000), whereas alfalfa leaves exposed to Hg^{2+} produce excess levels of both H_2O_2 and O_2^- (Zhou et al. 2008). This increase in ROS affects many other cellular functions by damaging nucleic acids and proteins, and by inducing lipid peroxidation thus modifying membrane integrity and permeability (Patra et al. 2004). In tomato, the production of ROS correlates with an increased activity of CAT, SOD, and PRX enzymes, in both roots and shoots (Cho and Park 2000).

1.3.3 Lead

Lead (Pb) is one of the most abundant heavy metals in both terrestrial and aquatic environments, predominantly arising from human activities such as mining, smelting, the use of fuels and explosives, and the disposal of Pb-enriched municipal sewage sludge. Together with Cd, Pb is also considered one of the most serious hazards to human health, since it is readily taken up by plants and therefore can easily enter the food chain. Pb toxicity causes similar symptoms to other heavy metals, namely growth inhibition, chlorosis, and (in the most severe cases) death.

Roots that absorb Pb respond by reducing their growth rate and changing their branching pattern. In *Picea abies*, the emergence and growth of secondary roots are particularly sensitive to Pb toxicity (Godbold and Kettner 1991). In maize, Pb perturbs the organization of the microtubule network of the root meristem, resulting in a shorter branching zone with more compact lateral roots emerging nearer to the root tips (Eun et al. 2000). The inhibition of root growth by Pb also affects nutrient uptake and nitrogen assimilation. For example, the enzymes nitrate reductase and glutamine synthetase are inhibited by Pb in *Cucumis sativus* and *Glycine max*, respectively (Sharma and Dubey 2005). Pb also nonspecifically blocks the uptake of other cations such as K, Ca, Mg, Mn, Zn, Cu, and Fe, probably by modifying the activity and permeability of membranes or binding them to ion carriers, making them unavailable for uptake and transport into the plant (Patra et al. 2004).

High concentrations of Pb cause a water deficit, reducing the transpiration rate, altering the osmotic pressure of the cell sap and the water potential of the xylem. These effects contribute to an overall negative change in the plant water status (Parys et al. 1998).

Pb interacts with –SH groups like many other heavy metals, but it can also interact with –COOH groups, inhibiting enzymes and altering protein conformation (Sharma and Dubey 2005). Pb can also displace metal cofactors from metalloenzymes, which includes Mn in the OEC and Mg in the chlorophyll porphyrin ring, thus interfering with photosynthesis and electron transport by reducing oxygen evolution and chlorophyll levels and altering the thylakoid membrane structure (Patra et al. 2004). Key chlorophyll biosynthesis enzymes are also strongly inhibited by Pb, as well as many enzymes in the Calvin cycle (e.g., RuBisCO, phosphoenol pyruvate carboxylase, and ribulose 5-phosphate kinase) thus reducing the rate and efficiency of CO_2 fixation (Sharma and Dubey 2005).

One unique effect of Pb is the disruption of the cell cycle by interfering with the alignment of microtubules on the mitotic spindle. This effect cannot be replicated with, e.g., Al and Cu, even at concentrations sufficient to inhibit root growth (Eun et al. 2000).

Pb is not a redox metal so cannot generate ROS directly, but oxidative stress is caused indirectly as shown by the increased lipid peroxidation in rice and pea plants exposed to the metal (Malecka et al. 2001). This is countered by the activation of antioxidant enzymes such as SOD and PRX, but whereas CAT activity increases in pea plants, it declines in rice, perhaps explaining in part why there is an increase in lipid peroxidation (Malecka et al. 2001, Verma and Dubey 2003). This complexity of antioxidant enzyme activity in plants under metal stress may reflect the presence of diverse isoforms which have different spatiotemporal expression profiles, different intracellular locations, and different environmental triggers for activation and inactivation (Scandalios 1990).

1.3.4 Chromium

Chromium (Cr) has received comparatively little attention from plant scientists perhaps because it is ubiquitous in the environment and, due to its complex electron chemistry, it exists in many oxidation states upon which its toxicity depends. Cr pollution results from human activities such as leather processing and finishing, the production of refractory steel, electroplating, wood preservation, and the manufacture of specialty chemicals and cleaning agents such as chromic acid. There is no evidence that Cr has a specific biological in plants, and its absorption involves the use of Fe, S, and P transporters and carriers; Cr thus competes with these essential nutrients for binding sites. Cr ions with different oxidation state appear to be absorbed by different mechanisms (Shanker et al. 2005). Cr stress inhibits germination in *Phaseolus vulgaris*, possibly by promoting the activity of proteases while suppressing the activity of amylases and perturbing the subsequent transport of sugars to the embryo axes (Zeid 2001). In adult plants, Cr toxicity inhibits shoot growth, reduces the number of leaves as well as the leaf area and biomass, reduces the productivity of crops, causes burns on the leaf margins and tips, and induces chlorosis and necrosis (Sharma and Sharma 1993; Singh 2001; Jain et al. 2000). Eventually, the global plant fitness is compromised, giving reduced plant biomass production and productivity, relevant aspects for crops and agronomy-important species.

A well-documented effect of Cr toxicity is the inhibition of primary root growth (observed as reduced root length) and the suppression of new lateral root primordia

(Prasad et al. 2001). The application of Cr inhibited root elongation in *Caesalpinia pulcherrima,* wheat, and *Vigna radiata* (Shanker et al. 2005) possibly by disrupting cell division through chromosomal damage (Panda and Choudhury 2005). Cr stress also induces changes in root morphology, increasing the number of root hairs and the relative proportion of pith and cortical tissue layers (Suseela et al. 2002). The negative effects of Cr on root growth and development combined with the tendency of Cr to compete with essential nutrients for uptake and transport means that Cr has a significant impact on nutrient acquisition. Although there is some variation depending on the plant species and tissue, Cr(VI) seems to have the most potent effect on the uptake of nutrients such as K, Mg, P, Fe, N, Zn, Cu, Mo, and Mn (Shanker et al. 2005). As well as reducing root growth and competing with these essential nutrients for uptake, Cr may also inhibit the activity of $H^+ATPases$ in the plasma membrane, which is required for proton export from the roots and hence acidification of the rhizosphere and the subsequent mobilization of metal ions. Inhibition would therefore result in a general reduction in nutrient bioavailability in the soil (Shanker et al. 2005).

The impact of Cr on plant water status in unknown, although Cr does induce the typical symptoms of water deficit and reduced transpiration, such as turgor loss, plasmolysis, and diminished tracheary vessel diameter (Shanker et al. 2005).

Both photosystems are inhibited by Cr(VI) although the mechanisms are still under investigation. Exposure to Cr(III) and Cr(VI) reduces the chlorophyll content of bean seedlings and wheat plants by displacing Mg from the chlorophyll molecule (Samantaray et al. 2001; Sharma and Sharma 1996). Cr stress also disrupts the ultrastructure of the chloroplast, particularly the arrangement of thykaloid membranes, probably reducing the size of the antenna complexes (Panda and Choudhury 2005; Shanker et al. 2005). ·

Cr can also inhibit certain enzymes in a species-dependent manner, e.g. nitrate reductase (Panda and Patra 2000) and root Fe(III) reductase (Barton et al. 2000), the latter affecting Fe nutrition in the plant. In mitochondria, Cr may hamper the electron transport interfering with the Cu and Fe ions contained in many electron-carrier proteins. The severe inhibition of mitochondrial cytochrome oxidation, for instance, could be due to the extreme susceptibility of complex III and IV to Cr(VI) (Dixit et al. 2002).

Finally, Cr shares the ability of other heavy metals to induce the formation of ROS in plant cells. Cr is not considered a redox metal, but studies have shown that it can participate in Fenton reactions (Panda and Choudhury 2005). Sorghum plants treated with either Cr(VI) or Cr(III) increased H_2O_2 content in roots and leaves, correlated with an increase in lipid peroxidation (Panda and Choudhury 2005; Shanker and Pathmanabhan 2004). Antioxidant enzyme activities are also modulated by Cr, apparently in a dose-dependent manner. For example, low levels of Cr induce SOD activity in pea plants, whereas higher concentrations inhibit both CAT and SOD (Dixit et al. 2002; Jain et al. 2000).

1.3.5 Arsenic

Arsenic (As) is a profoundly toxic heavy metalloid that originates from both geogenic sources and anthropogenic activities such as mining, the combustion of fossil fuels, and use of As-based pesticides and wood preservatives (Tu and Ma 2005). It is widely distributed in the environment and recognized as a significant threat to human health. The chemistry of As in the soil is complex because it can be present in both organic and inorganic forms, but most As is present as the oxidized mineral arsenate, AsO_4^{3-} As(V), and its reduced form arsenite, AsO_3^{3-} As(III). The bioavailability of As depends on the soil characteristics, including its redox potential, pH, and composition, the presence of other minerals (particularly Fe and Al oxides and hydroxides), and the abundance of microbes that can reduce As(V) to As(III) (Smith et al. 2010). Arsenate is chemically similar to phosphate and it is probably taken up into many plants via phosphate transporters (Pigna et al. 2009). In contrast, arsenite is more abundant and mobile in soils with a low redox potential, and is thought to be acquired via aquaporin transporters in the plasma membrane of root cells (Vromman et al. 2011).

As interferes with cell metabolism by reacting with –SH groups on proteins and replacing phosphate, and inhibits plant growth (Tu and Ma 2005). The symptoms of As toxicity include poor seed germination and profound growth inhibition (Smith et al. 2010). In wheat seeds, for example, germination is considerably affected by both arsenite and arsenate, probably reflecting the inhibition of both α- and β-amylase (Liu et al. 2005). Maize plants treated with toxic concentration of As(V) and As(III) produced stunted roots that were thicker and stiffer than normal, and that had a significantly lower mitotic index; micronuclei and chromosome aberrations were also observed in the root meristems (Duquesnoy et al. 2010). In some species, the effect of As on root growth depends on its concentration. For example, root growth in *Artemisia annua* is stimulated at low As concentrations but inhibited at higher concentrations (Rai et al. 2011b).

The reduction in root growth combined with changes in the selectivity and permeability of cell membranes prevent the uptake of water and nutrients resulting in water imbalance and nutrient deficiency, the severity depending on the species (Paivoke and Simola 2001). For example, As significantly increases the accumulation of N, P, K, Ca, and Mg in the shoots of hydroponically grown *Phaseulus vulgaris* plants (Carbonell-Barrachina et al. 1997), but reduces the uptake of macronutrients such as K, Ca, and Mg, and micronutrients such as B, Cu, Mn, and Zn, in tomato plants (Carbonell-Barrachina et al. 1994). Similarly, arsenite reduces the uptake of Si, Mn, Zn, Cu, P, and K in rice plants and the translocation of these minerals to the shoot, possibly by interacting with the –SH groups of transporters (Hoffmann and Schenk 2011). Interestingly, in some hyperaccumulator species, such as *Pteris vittata*, low levels of arsenate stimulate phosphate accumulation in the fronds and significantly enhance growth (Tu and Ma 2005). The water content, water potential, and stomatal conductance of *Atriplex atacamensis* (Phil) leaves and roots were significantly reduced after prolonged exposure to As (Vromman et al. 2011).

Following absorption, As is thought to interfere with essential phosphate metabolism because the corresponding enzymes can also reduce As(V) to As(III) (Smith et al. 2010). Moreover, As(V) can be reduced *nonenzymatically* by glutathione (at least in vitro; Meharg and Hartley-Whitaker 2002) which is abundant in plants. Although As is not redox-active, it can stimulate the production of ROS through the conversion of arsenate to arsenite (Meharg and Hartley-Whitaker 2002), and can thus induce lipid peroxidation and cellular damages (Gunes et al. 2009). Maize leaves and roots exposed to As(V) produce antioxidant enzymes such as APX in response to the oxidative stress, whereas SOD activity declines. Conversely, higher levels of CAT activity were measured in maize shoots and roots exposed to high concentrations of As(III) (Duquesnoy et al. 2010). In *Bacopa monnieri* plants exposed to moderate levels of As, the activities of GSR, SOD, GPX, APX, and CAT were stimulated in a differential but coordinated manner in the leaves and roots, presumably representing a global response to As toxicity (Mishra et al. 2011). *Artemisia annua* plants treated with As showed a dose-dependent increase in the activities of SOD, APX, GSR, and GPX followed by a gradual decline at higher concentrations, again suggesting a coordinated response to the oxidative stress caused by As toxicity (Rai et al. 2011b).

1.4 Essential Metal Ions: Nickel, Copper, Iron, Manganese, Zinc, and Selenium

1.4.1 Nickel

Nickel (Ni) is abundant in rocks as a free metal and as a complex with other metal ions such as Fe. Like other heavy metals, anthropogenic activities such as mining, smelting, burning fossil fuels, vehicle emissions, waste disposal, electroplating, and the manufacture and disposal of batteries contribute to the release of Ni into the environment (Alloway 1995; Salt et al. 2000). Like Cr, Ni exists in many oxidation states that complicate the investigation of toxicity mechanisms in plants. However, Ni^{2+} is the prevalent oxidation state in soils because it is stable over a wide range of pH and redox conditions (Yusuf et al. 2011).

Unlike the metals discussed above, Ni is an essential micronutrient because it is required as a cofactor in enzymes such as urease, where it usually coordinates with cysteine residues (Dixon et al. 1980). Ni deficiency therefore reduces urease activity, disrupts nitrogen metabolism, and leads to the accumulation of toxic amounts of urea, which manifests as chlorosis and necrosis (Yusuf et al. 2011). These effects are particularly severe in species that develop symbiotic relationships with nitrogen-fixing bacteria, because amino acid metabolism and the ornithine cycle are also compromised (Eskew et al. 1983). Low levels of Ni thus promote growth and development in many crops, including oilseed rape, cotton, sweet pepper, tomato, and potato (Gerendas and Sattelmacher 1999; Welch 1981).

Although Ni is an essential nutrient, excess amounts are toxic in many species and the effects are already apparent during germination in species such as pigeon pea, maize, wheat, and *B. juncea* (Rao and Sresty 2000; Bhardwaj et al. 2007; Gajewska and Sklodowska 2008; Sharma et al. 2008). Later in development, the inhibition of root growth is a prevalent symptom of Ni toxicity, as seen in *B. juncea* plants and wheat seedlings (Alam et al. 2007; Gajewska et al. 2006). The uptake of nutrients is also affected by Ni excess, and its chemical similarity to nutrients such as Ca, Mg, Mn, Fe, Cu, and Zn suggests that Ni may compete with these minerals for uptake and subsequent utilization (Chen et al. 2009). Excess Ni may therefore induce deficiency symptoms for other nutrients, e.g., in barley plants where toxic levels of Ni reduce the absorption of Ca, Fe, K, Mg, Mn, P and Zn (Brune and Deitz 1995). Excess Ni also reduces the level of nitrogen in the leaves and roots of *Cicer arietinum* and *Vigna radiata* plants (Athar and Ahmad 2002). Ni exposure reduces the phosphorus content of *Helinathus annus* and *Hyptis suaveolens* plants (Pillay et al. 1996).

Like other heavy metals, Ni also disrupts the water balance in plants, perhaps reflecting the cumulative effects of Ni toxicity. Indeed, Ni treatment reduces the transpiration rate, leaf growth, and the leaf blade area in wheat (Bishnoi et al. 1993; Chen et al. 2009), and increases the level of endogenous ABA in *Brassica oleracea* leaves, the plant hormone that promotes stomatal closure (Molas 1997).

Ni has a substantial impact on photosynthesis because it disrupts the thylakoid membranes and reduces the chlorophyll content (Molas 2002; Ahmad et al. 2007; Alam et al. 2007). Like other metals, Ni can displace Mg from chlorophyll and enzymes such as RuBisCO that contain Mg ion as cofactor (Yusuf et al. 2011). Moreover, both PSI and the PSII appear to be sensitive to Ni in *Spinacea oleracea*, where the analysis of submembrane fractions showed that Ni^{2+} strongly inhibits oxygen evolution by depleting the extrinsic 16 and 2 kDa polypeptides associated with the OEC (Boisvert et al. 2007).

Unlike Fe and Cu (see below), Ni is not a redox-active metal and cannot generate ROS directly, yet the presence of excess Ni nevertheless induces the formation of superoxide anions, hydroxyl radicals, and hydrogen peroxide in many species, including *Alyssum bertolonii* and wheat (Boominathan and Doran 2002; Gajewska and Sklodowska 2007). Interestingly, the prolonged presence of these ROS does not increase the amount of lipid peroxidation in wheat, perhaps due to the concomitant increase in APX and GPX activities (Gajewska and Sklodowska 2007). In a different experiment, the treatment of *Triticum durum* with Ni^{2+} resulted in a significant increase in membrane lipid peroxidation, along with higher levels of H_2O_2 and O_2^- (Hao et al. 2006). Similarly, H_2O_2 levels rose significantly following the exposure of both *Alyssum bertoloni* and *Nicotiana tabacum* to Ni, although there was little oxidative damage in *A. bertolonii* roots reflecting the much higher endogenous activities of CAT and SOD in this species (Boominathan and Doran 2002). The induction or repression of antioxidant enzymes is species dependent and also reflects the magnitude of the stress. For example, while in *A. bertolonii*, SOD, CAT, and APX activities decline in response to Ni, the opposite pattern is observed in wheat and maize (Baccouch et al. 2001; Ga-jewska et al. 2006).

1.4.2 Copper

Copper (Cu) is an essential nutrient that acts as a structural component in regulatory proteins, as a redox component in chloroplast and mitochondrial electron transport, and as a cofactor in enzymes such as Cu-SOD, cytochrome oxidase, plastocyanin, and laccase, therefore participating in a variety of metabolic processes, such as hormone signaling, cell wall metabolism, and stress response. Cu deficiency symptoms include chlorosis and necrosis at the leaf tip, together with leaf twisting and malformation, reflecting the impairment of photosynthetic electron transfer, the loss of essential pigments, and the degeneration of thykaloids.

Cu in plants exists in two oxidation states, Cu^{2+} and Cu^+, and *redox cycling* between these states produces hydroxyl radicals (Li et al. 2002). Moreover, since Cu is a redox-active transition metal it can generate ROS directly via the Fenton or Haber–Weiss reactions (Halliwell and Gutteridge 1984), catalyzing the formation of hydroxyl radicals (OH') via non-enzymatic chemical reactions between H_2O_2 and the superoxide anion (O_2^-) (see Fig. 1.1). This enhanced capacity to produce ROS is the primary mechanism of Cu toxicity.

Further visible symptoms of Cu toxicity include stunted growth and reduced initiation and development of lateral roots. Nitrogen metabolism and fixation are disrupted in *Glycine max* plants exposed to excess Cu, whereas nitrate and free amino acid levels become depleted in similarly treated *Vitis vinifera* plants (Llorens et al. 2000).

One of the most potent effects of Cu toxicity is to inhibit oxygen evolution, accompanied by a significant reduction in photosynthetic yield, which may reflect a specific interaction between Cu ions and Tyr_Z and Tyr_D on the D2 protein of PSII (Sabat 1996; Maksymiec and Baszynski 1999). The photosynthetic machinery is strongly inhibited by excess Cu, resulting in the degradation of stromal lamellae, the loss of grana stacking, and an increase in the number and size of plastoglobules (Yruela 2005). The extrinsic proteins of the OEC (PsbO, PsbP and PsbQ) are degraded in the presence of excess Cu (Yruela 2005) and the redox state of cytochrome b_{559} is compromised (Roncel et al. 2001). Important enzymes of the Calvin cycle are also inhibited by Cu, including RuBisCO and phosphoenol pyruvate caboxylase (Balsberg Pahlsson 1989). Cu stress increases susceptibility to photoinhibition in both isolated thylakoids and intact leaves, due to the Cu-induced reduction of chlorophyll content (Pätsikkä et al. 2002).

1.4.3 Iron

Iron (Fe) is an essential nutrient in plants, with crucial roles in processes such as photosynthetic electron transport, oxidative stress tolerance, mitochondrial respiration, nitrogen fixation, hormone synthesis and organelle maintenance (Hänsch and Mendel 2009). It exists in soils as both Fe^{3+} and Fe^{2+}, although only the latter is soluble and suitable for absorption by plants. Fe geochemistry is influenced by soil characteristics such as pH, organic matter content, and oxygen levels. Fe^{3+} is reduced to Fe^{2+} by soil microorganisms, root exudates, and chemical reactions in the soil. An interesting feature of Fe toxicity is that it is greatly dependent on the soil type; it is often linked to P and Zn deficiency, water logging, and anoxic conditions (Ponnamperuma et al. 1967).

Fe is a highly reactive redox metal that produces large amounts of hydrogen peroxide and superoxide during the reduction of molecular oxygen. Therefore, excess Fe induces the formation of hydroxyl radicals that can damage many targets, including DNA, proteins, lipids, and sugars. A typical visual symptom of iron toxicity in rice is the *bronzing* of leaves due to the accumulation of oxidized polyphenols (Becker and Asch 2005). In *Nicotiana plumbaginifolia* and pea plants, Fe toxicity induces the formation of brown necrotic spots covering the whole leaf surface (Kampfenkel et al. 1995). In wetland plants, iron oxyhydroxide deposits (*iron plaques*) may form on the roots in Fe-rich soils. These deposits reduce further Fe absorption, thus constituting a protective mechanism against Fe toxicity, but they may also sequester nutrients such as phosphate and therefore result in deficiency symptoms (Batty and Younger 2003). Excess Fe reduces water transpiration and photosynthetic activity (Kampfenkel et al. 1995; Adamski et al. 2011), which manifests, for instance, as a sharp decline in the chlorophyll content of potato leaves (Chatterjee et al. 2006) together with a loss of thylakoid membrane integrity (Adamski et al. 2011), and as a reduction in CO_2 fixation and starch accumulation in *N. plumbaginifolia* plants (Kampfenkel et al. 1995). Fe stress in sweet potatoes inhibits the reduction of plastoquinone but appears not to affect electron flux from plastoquinone to the final electron acceptor (Adamski et al. 2011).

Because Fe participates in the Haber–Weiss and Fenton reactions, one of the main toxicity mechanisms is the direct formation of ROS and the induction of oxidative stress, which has been documented in *N. plumbaginifolia,* rice, sunflowers, and soybean (Kamplenken et al. 1995; Fang et al. 2001). As a response to the increased oxidative stress, activity of APX and GSR were increased by Fe^{2+} excess in rice, as well as the amount of free-radical scavengers, such as mannitol and reduced GS (Fang et al. 2001). Application of Fe^{2+} ions was also found to induce peroxidase activity in rice leaves, which could be mediated by *de novo* synthesis of the enzyme at transcriptional level (Peng et al. 1996). CAT and APX were shown to be induced in *N. plumbaginifolia* plants exposed to excess Fe (Kampfenkel et al. 1995).

1.4.4 Manganese

Manganese (Mn) acts as cofactor in many enzymes, including Mn-superoxide dismutase, catalase, pyruvate carboxylase, phosphoenol pyruvate carboxykinase, malic enzyme, and isocitrate lyase (Hänsch and Mendel 2009). It also has a critical role in oxygen evolution because four Mn atoms are required in the OEC subunits of PSII. The oxidation state and bioavailability of Mn is strongly dependent on soil pH, with the more-soluble Mn(II) form becoming more abundant below pH 5.5 (thus risking Mn toxicity) and the less-soluble manganic forms Mn(III), Mn(IV), and Mn(VII) becoming more abundant above pH 6.5 (thus risking Mn deficiency). Mn utilization and toxicity is therefore exquisitely sensitive to fertilizer applications, particularly ammonia-based chemicals that cause soil acidification (Dučic and Polle 2005).

In *Zea mays* for instance, Mn deficiency restricted the uptake and transport of NO_3^-, inhibited the activity of enzymes related to N-metabolism, such as nitrate reductase, glutamine synthetase, and glutamic-oxaloace transaminase. Mn deficiency also promotes glutamate dehydrogenase activity, reduces chlorophyll and protein synthesis, and thus inhibits growth and development (Gong et al. 2011).

Excess Mn, for instance in low-drained and acidic soils, is toxic in most plant species, inducing general symptoms such as stunting, chlorosis, crinkled leaves, brown necrotic lesions, and death in the most severe cases (Dučic and Polle 2005). Pea plants exposed to excess Mn had lower root and shoot biomass, lower chlorophyll and carotenoid contents, and lower glutamine synthetase and glutamate synthase activities than controls (Gangwar et al. 2011). In *Vigna radiata* leaves, Mn treatment caused a progressive reduction in the total carotenoid, total chlorophyll and chlorophyll *a* contents, and inhibited the Hill activity of isolated chloroplasts, thus reducing the rates of photosynthesis and of CO_2 uptake (Sinha et al. 2002). The decline in photosynthetic activity following exposure to excess Mn also reflects the production of ROS such as H_2O_2 and O_2^- (Gangwar et al. 2011; Shi and Zhu 2008), which cause lipid preoxidation in the thylakoid membranes and damage enzymes such as RuBisCO (Subrahmanyam and Rathore 2001). In plants exposed to high Mn levels, the activity of SOD, PRX, APX, DHAR and GSR is increased. Excess Mn can also cause deficiencies for other nutrients such as Fe, Mg, Zn and Ca, although the mechanism is unclear (Shi and Zhu 2008).

1.4.5 Zinc

Zinc (Zn) is an essential element and participates in many processes of plant life, such as enzyme activation, metabolism of proteins and carbohydrates, lipids, and nucleic acids. Zn is a cofactor in many plant enzymes with important roles in primary metabolism (e.g., alcohol dehydrogenase, glutamic dehydrogenase, carbonic anhydrase, enzymes involved in electron transport, and antioxidant enzymes) and is also an integral component of several transcription factors (e.g., zinc finger transcription factors) (Chang et al. 2005). Zn deficiency initially manifests as a reduction in internodal growth, which reduces stem length and causes plants to acquire rosette-like *habitus*, and in later stages leaves may develop deficiency symptoms such as chlorosis and necrotic spots (Sharma 2006).

Zn is usually abundant in the mineral component of soils and is present as sulfide, sulfate, oxide, carbonate, phosphate, and silicate, e.g., sphalerite, zincosite, gahnite, smithsonite, hopeite, and willemite (Broadley et al. 2007). Zn levels in soils have also increased through human activities such as mining, smelting, limestone topping, burning fossil fuels, and the use of phosphate-based fertilizers (Nriagu 1996). Under physiological conditions, the relatively stable Zn^{2+} redox state is prevalent in soils although this depends on the soil type, clay and mineral content, moisture content, weathering rates, organic matter content, and microbial populations. The most important parameter is soil pH; Zn is more readily adsorbed on cation exchange sites at high pH, while it is more soluble in acidic soils with low levels of soluble organic matter and these conditions favor Zn toxicity (Broadley et al. 2007).

The initial symptoms of Zn toxicity are chlorosis and even reddening of the leaves in severe cases, due to anthocyanin production (Fontes and Cox 1995). This is followed by the appearance of necrotic brown spots on the leaves of some species, accompanied by stunting and reduced yield (Harmens et al. 1993; Broadley et al. 2007). Zn toxicity also inhibits primary root growth and the emergence of lateral roots (Ren et al. 1993). High levels of Zn can displace Mg from the OEC water splitting site of PSII, thus inhibiting both photosystems and the electron transport chain, as seen in Zn-treated *Phaseolus vulgaris* plants (Van Assche and Clijsters 1986). In *Spinacea oleracea*, plastidial ATP synthesis is also inhibited by Zn toxicity (Teige et al. 1990). Zn is a non-redox metal but it can generate ROS indirectly, leading to defense responses including the induction of antioxidant enzymes such as SOD, CAT, and GPX (Prasad et al. 1999; Chang et al. 2005). The oxidative burst induced by Zn toxicity could also be responsible for the cell death observed in rice root cells, since the application of exogenous ROS scavengers was able to increase cell viability; this result points to a relationship between Zn toxicity and programmed cell death (Chang et al. 2005).

1.4.6 Selenium

Although it has a relatively low density (4.82 g cm^{-3}) and according to the periodic table, it is a non-metal, Se is considered in this article because it shares many biological properties with other minerals (e.g., it exists in the soil in multiple forms and can induce toxicity symptoms depending on availability and abundance). In aerobic soils, inorganic Se is present in numerous oxidation states, the most common being selenite [SeO_3^{2-}, Se(IV)] and selenate [SeO_4^{2-}, Se(VI)], which are the most soluble and the most toxic forms. Elemental selenium (Se^0), which is more prevalent under anaerobic conditions, is insoluble and biologically inert. Inorganic Se is released naturally during the erosion and the leaching of seleniferous minerals, and many human activities also produce inorganic Se, including mining, burning fossil fuels, and glass manufacturing (Di Gregorio et al. 2005). Se is a nutrient, essential in traces for bacteria, animals, and algae, being a component of few enzymes, such as the glutathione peroxidase, in which it is incorporated as Se-cystein, encoded by the opal codon UGA (Fu et al. 2002). The status of Se as a micronutrient in higher plants remains controversial. Se stimulates the growth of Se-hyperaccumulators such as *Astragalus pectinatus* (Trelease and Trelease 1939), but true seleno-proteins similar to those found in microbes, animals, and algae have not yet been identified (Fu et al. 2002). Se can also regulate the water status of plants subjected to drought stress, increasing the water uptake capacity of roots and inhibiting the stress-induced accumulation of proline (Kuznestov et al. 2003). At low concentrations, Se behaves as an antioxidant in *Lolium perenne*, inhibiting lipid peroxidation and enhancing the activity of GPX (Hartikainen et al. 2000). The foliar application of Se to heat-stressed sorghum plants alleviates oxidative stress by enhancing the antioxidative cycle (Djana-guiraman et al. 2010).

Both selenate and selenite are readily absorbed by the roots of many plant species and are efficiently distributed to other tissues. Here, cellular metabolism converts them into Se-metabolites, which act as analogs of organic sulfur compounds and interfere with the metabolic processes in which these sulfur compounds normally participate. Moreover, the sulfur-containing amino acids cysteine and methionine are replaced by seleno-cysteine and seleno-methionine, which become incorporated into proteins, leading to significant alterations in protein function and structure due to the differences in size and ionization properties between Se and sulfur (Brown and Shrift 1982). High levels of Se trigger a range of toxicity symptoms including stunting, chlorosis, drying of leaves, aberrant protein metabolism, and eventually death. Symptoms vary according to (i) the age of the plant (older plants are more resistant) (ii) the assimilation characteristics of the plant (certain species hyperaccumulate Se); and (iii) the availability of sulfates, which compete with Se and mitigate its toxicity (Terry et al. 2000). Proteomic analysis in rice showed that Se toxicity led to a gradual decline in the chloroplast enzymes involved in the redox cycle (ROS scavenging system) and a corresponding gradual increase in the abundance of ROS and damage to the photosynthetic apparatus, in particular the chlorophyll a–b binding proteins and RuBisCO (Wang et al. 2012). This inhibition of photosynthesis combined with the impact of seleno-cysteine and seleno-methionine on protein synthesis and metabolism could therefore explain the reduced growth of rice seedlings caused by excess Se. Finally, it appears that excess Se can also imbalance the uptake of other nutrients, e.g., increasing the intracellular concentration of Ca, Fe, Cu, Mn, and Zn but reducing P levels in *Trifolium repens* (Wu and Huang 1992). Furthermore, due to the chemical similarity between S and Se, SeO_4^{2-} and SO_4^{2-} probably compete for absorption and transport (Grant et al. 2011), reducing the amount of SO_4^{2-} absorbed by the roots (Leggett and Epstein 1956).

References

Adamski JM, Peters JA, Danieloski R, Bacarin MA (2011) Excess iron-induced changes in the photosynthetic characteristics of sweet potato. J Plant Physiol 168:2056–2062

Ahmad MSA, Hussain M, Saddiq R, Alvi AK (2007) Mungbean: a nickel indicator, accumulator or excluder? Bull Environ Contam Toxicol 78:319–324

Alam MM, Hayat S, Ali B, Ahmad A (2007) Effect of 28- homobrassinolide treatment on nickel toxicity in *Brassica juncea*. Photosynthetica 45:139–142

Alloway BJ (1995) Heavy metal in soils. Blackie Academic and Professional, London

Astolfi S, Zuchi S, Passera C (2004) Role of sulphur availability on cadmium-induced changes of nitrogen and sulphur metabolism in maize (*Zea mays* L.) leaves. J Plant Physiol 161:795–802

Athar R, Ahmad M (2002) Heavy metal toxicity in legume-microsymbiont system. J Plant Nutr 25:369–386

Baccouch S, Chaoui A, El Ferjani E (2001) Nickel toxicity induces oxidative damage in *Zea mays* roots. J Plant Nutr 24:1085–1097

Balestrasse KB, Benavides MP, Gallego SM, Tomaro ML (2003) Effect of cadmium stress on nitrogen metabolism in nodules and roots of soybean plants. Func Plant Biol 30:57–64

Balsberg Pahlsson AM (1989) Toxicity of heavy metals (Zn, Cu, Cd, Pb) to vascular plants. Water Air Soil Pollut 47:287–319

Barton LL, Johnson GV, O'Nan AG, Wagener BM (2000) Inhibition of ferric chelate reductase in alfalfa roots by cobalt, nickel, chromium, and copper. J Plant Nutr 23:1833–1845

Batty LC, Younger PL (2003) Effects of external iron concentration upon seedling growth and uptake of Fe and phosphate by the common reed, *Phragmites australis* (Cav.) trin ex. steudel. Ann Bot 92:801–806

Becker M, Asch F (2005) Iron toxicity in rice—conditions and management concepts. J Plant Nutr Soil Sci 168:558–573

Benavides MP, Gallego SM, Tomaro ML (2005) Cadmium toxicity in plants. Braz J Plant Physiol 17:21–34

Bernier M, Carpentier R (1995) The action of mercury on the binding of the extrinsic polypeptides associated with the water oxidizing complex of photosystem II. FEBS Lett 360:251–254

Bestwick CS, Brown IR, Mansfield JW (1998) Localized changes in peroxidase activity accompany hydrogen peroxide generation during the development of a non-host hypersensitive reaction in lettuce. Plant Physiol 118:1067–1078

Bhardwaj R, Arora N, Sharma P, Arora HK (2007) Effects of 28-homobrassinolide on seedling growth, lipid peroxidation and antioxidative enzyme activities under nickel stress in seedlings of *Zea mays* (L). Asian J Plant Sci 6:765–772

Bishnoi NR, Sheoran IS, Singh R (1993) Influence of cadmium and nickel on photosynthesis and water relations in wheat leaves of differential insertion levels. Photosynthetica 28:473–479

Boisvert S, Joly D, Leclerc S, Govindachary S, Harnois J, Carpentier R (2007) Inhibition of the oxygen-evolving complex of photosystem II and depletion of extrinsic polypeptides by nickel. Biometals 20:879–889

Boominathan R, Doran PM (2002) Ni-induced oxidative stress in roots of the Ni hyperaccumulator, *Alyssum bertolonii*. New Phytol 156:205–215

Boussama N, Ouariti O, Suzuki A, Ghorbel MH (1999) Cd-stress on nitrogen assimilation. J Plant Physiol 155:310–317

Broadley MR, White PJ, Hammond JP, Zelko I, Lux A (2007) Zinc in plants. New Phytol 173:677–702

Brown TA, Shrift A (1982) Selenium: toxicity and tolerance in higher plants. Biol Rev 57:59–84

Brune A, Deitz KJ (1995) A comparative analysis of element composition of roots and leaves of barley seedlings grown in the presence of toxic cadmium, molybdenum, nickel and zinc concentrations. J Plant Nutr 18:853–868

Carbonell-Barrachina A, Burlò-Carbonell F, Mataix-Beneyto J (1994) Effect of arsenite on the concentration of micronutrients in tomato plants grown in hydroponic culture. J Plant Nutr 17:1887–1903

Carbonell-Barrachina A, Burlò-Carbonell F, Mataix-Beneyto J (1997) Effect of sodium arsenite and sodium chloride on bean plant nutrition (macronutrients). J Plant Nutr 20:1617–1633

Chaffei C, Pageau K, Suzuki A, Gouia H, Ghorbel HM, Mascalaux-Daubresse C (2004) Cadmium toxicity induced changes in nitrogen management in *Lycopersicon esculentum* leading to a metabolic safeguard through an amino acid storage strategy. Plant Cell Physiol 45:1681–1693

Chang HB, Lin CW, Huang HJ (2005) Zinc-induced cell death in rice (*Oryza sativa* L.) roots. Plant Growth Regul 46:261–266

Chatterjee C, Gopal R, Dube BK (2006) Impact of iron stress on biomass, yield, metabolism and quality of potato (*Solanum tuberosum* L). Sci Hortic 108:1–6

Chen C, Huang D, Liu J (2009) Functions and toxicity of nickel in plants: recent advances and future prospects. Clean 37:304–313

Cho UH, Park JO (2000) Mercury-induced oxidative stress in tomato seedlings. Plant Sci 156:1–9

Christophersen HM, Smith SE, Pope S, Smith FA (2009) No evidence for competition between arsenate and phosphate for uptake from soil by medic or barley. Env Int 35:485–490

Clemens S (2006) Evolution and function of phytochelatin synthases. J Plant Physiol 163: 319–332

DalCorso G, Farinati S, Maistri S, Furini A (2008) How plants cope with cadmium: staking all on metabolism and gene expression. J Integr Plant Biol 50:1268–1280

DalCorso G, Farinati S, Furini A (2010) Regulatory networks of cadmium stress in plants. Plant Signal Behav 5:663–667

Dat JF, Vandenabeele S, Vranovà E, Van Montagu M, Inzè D, Van Breusegem F (2000) Dual action of the active oxygen species during plant stress responses. Cell Mol Life Sci 57: 779–795

Di Gregorio S, Lampis S, Vallini G (2005) Selenite precipitation by a rhizospheric strain of *Stenotrophomonas* sp. isolated from the root system of *Astragalus bisulcatus*: a biotechnological perspective. Environ Int 31:233–241

Dixit V, Pandey V, Shyam R (2002) Chromium ions inactivate electron transport and enhance superoxide generation in vivo in pea (*Pisum sativum* L. cv: Azad) root mitochondria. Plant Cell Environ 25:687–693

Dixon H, Hinds JA, Fihelly AK, Gozala C, Winzor DJ, Blakeley RL, Zerner B (1980) Jack bean urease (EC 3.5.1.5). IV. The molecular size and mechanism of inhibition by hydroxamic acids. Spectrophotometric fixation of enzymes with reversible inhibitors. Can J Biochem 58:1323–1334

Djanaguiraman M, Prasad PV, Seppanen M (2010) Selenium protects sorghum leaves from oxidative damage under high temperature stress by enhancing antioxidant defense system. Plant Physiol Biochem 48:999–1007

Dučic T, Polle A (2005) Transport and detoxification of manganese and copper in plants. Braz J Plant Physiol 17:103–112

Duquesnoy I, Champeau GM, Evray G, Ledoigt G, Piquet-Pissaloux A (2010) Enzymatic adaptations to arsenic-induced oxidative stress in Zea mays and genotoxic effect of arsenic in root tips of *Vicia faba* and *Zea mays*. C R Biol 333:814–824

Eskew DL, Welch RM, Cary EE (1983) Nickel: an essential micronutrient for legumes and possibly all higher plants. Science 222:691–693

Eun SO, Yon HS, Lee Y (2000) Lead distrurbs microtubule organization in the root meristem of *Zea mays*. Physiol Plant 110:357–365

Fang WC, Wang JW, Lin CC, Kao CH (2001) Iron induction of lipid peroxidation and effects on antioxidative enzyme activities in rice leaves. Plant Growth Regul 35:75–80

Farinati S, DalCorso G, Varotto S, Furini A (2010) The *Brassica juncea* BjCdR15, an ortholog of *Arabidopsis* TGA3, is a regulator of cadmium uptake, transport and accumulation in shoots and confers cadmium tolerance in transgenic plants. New Phytol 185:964–978

Fontes RLF, Cox FR (1995) Effects of sulfur supply on soybean plants exposed to zinc toxicity. J Plant Nut 18:1893–1906

Fu LH, Wang XF, Eyal Y, She YM, Donald LJ, Standing KG, Ben-Hayyim G (2002) A selenoprotein in the plant kingdom. Mass spectrometry confirms that an opal codon (UGA) encodes selenocysteine in *Chlamydomonas reinhardtii* gluththione peroxidase. J Biol Chem 277:25983–25991

Führs H, Götze S, Specht A, Erban A, Gallien S, Heintz D, Van Dorsselaer A, Kopka J, Braun HP, Horst WJ (2009) Characterization of leaf apoplastic peroxidases and metabolites in *Vigna unguiculata* in response to toxic manganese supply and silicon. J Exp Bot 60:1663–1678

Gabbrielli R, Pandolfini T (1984) Effect of Mg^{2+} and Ca^{2+} on the response to nickel toxicity in a serpentine endemic and nickel accumulating species. Physiol Plant 62:540–544

Gajewska E, Sklodowska M (2007) Effect of nickel on ROS content and antioxidative enzyme activities in wheat leaves. Biometals 20:27–36

Gajewska E, Sklodowska M (2008) Differential biochemical responses of wheat shoots and roots to nickel stress: antioxidative reactions and proline accumulation. Plant Growth Regul 54:179–188

Gajewska E, Sklodowska M, Slaba M, Mazur J (2006) Effect of nickel on antioxidative enzyme activities, proline and chlorophyll content in wheat shoots. Biol Plant 50:653–659

Gangwar S, Singh VP, Maurya JN (2011) Responses of *Pisum sativum* L. to exogenous indole acetic acid application under manganese toxicity. Bull Environ Contam Toxicol 86:605–609

Gerendas J, Sattelmacher B (1999) Influence of Ni supply on growth and nitrogen metabolism of *Brassica napus* L. grown with NH_4NO_3 or urea as N source. Ann Bot 83:65–71

Godbold DL, Kettner C (1991) Lead influences root growth and mineral nutrition of *Picea abies* seedlings. J Plant Physiol 139:95–99

Gong X, Qu C, Liu C, Hong M, Wang L, Hong F (2011) Effects of manganese deficiency and added cerium on nitrogen metabolism of maize. Biol Trace Elem Res 144:1240–1250

Grant K, Carey NM, Mendoza M, Schulze J, Pilon M, Pilon-Smits EA, van Hoewyk D (2011) Adenosine 5'-phosphosulfate reductase (APR2) mutation in *Arabidopsis* implicates glutathione deficiency in selenate toxicity. Biochem J 438:325–335

Gunes A, Pilbeam DJ, Inal A (2009) Effect of arsenic phosphorus interaction on arsenic-induced oxidative stress in chickpea plants. Plant Soil 314:211–220

Hall JL (2002) Cellular mechanisms for heavy metal detoxification and tolerance. J Exp Bot 53:1–11

Halliwell B, Gutteridge JMC (1984) Oxygen toxicity, oxygen radicals, transition metals and disease. Biochem J 219:1–14

Han FX, Su Y, Monts DL, Waggoner AC, Plodinec JM (2006) Binding, distribution, and plant uptake of mercury in a soil from Oak Ridge, Tennessee, USA. Sci Total Environ 368:753–768

Hänsch R, Mendel RR (2009) Physiological functions of mineral micronutrients (Cu, Zn, Mn, Fe, Ni, Mo, B Cl). Curr Opin Plant Biol 12:259–266

Hao F, Wang X, Chen J (2006) Involvement of plasma-membrane NADPH oxidase in nickel-induced oxidative stress in roots of wheat seedlings. Plant Sci 170:151–158

Harmens H, Gusmao NGCPB, Hartog DPR, Verkeij JAC, Ernst WHO (1993) Uptake and transport of zinc in zinc-sensitive and zinc-tolerant *Silene vulgaris*. J Plant Physiol 141:309–315

Hartikainen H, Xue T, Piironen V (2000) Selenium as an anti-oxidant and pro-oxidant in ryegrass. Plant Soil 225:193–200

Hoffmann H, Schenk MK (2011) Arsenite toxicity and uptake rate of rice (*Oryza sativa* L.) in vivo. Environ Pollut 159:2398–2404

Horst WJ, Fecht M, Naumann A, Wissemeier AH, Maier P (1999) Physiology of manganese toxicity and tolerance in *Vigna unguiculata* (L.) Walp. J Plant Nut Soil Sci 162:263–274

Jain R, Srivastava S, Madan VK, Jain R (2000) Influence of chromium on growth and cell division of sugarcane. Indian J Plant Physiol 5:228–231

Kampfenkel K, Van Montagu M, Inzé D (1995) Effects of iron excess on *Nicotiana plumbaginifolia* plants—implications to oxidative stress. Plant Physiol 107:725–735

Kuznetsov VV, Kholodova VP, Kuznetsov VV, Yagodin BA (2003) Selenium regulates the water status of plants exposed to drought. Doklady Biol Sci 390:266–268

Lane TW, Morel FM (2000) A biological function for cadmium in marine diatoms. Proc Natl Acad Sci U S A 97:4627–4631

Le Bot J, Goss MJ, Carvalho MJGPR, Beusichem ML, Kirkby EA (1990) The significance of the magnesium to manganese ratio in plant tissues for growth and alleviation of manganese toxicity in tomato (*Lycopersicon esculentum*) and wheat (*Triticum aestivum*) plants. Plant Soil 124:205–210

Leggett JE, Epstein E (1956) Kinetics of sulfate absorption by barley roots. Plant Physiol 31: 222–226

Li Y, Seacat A, Kuppusamy P, Zweier JL, Yager JD, Trush MA (2002) Copper redox-dependent activation of 2-tert-butyl(1,4)hydroquinone: formation of reactive oxygen species and induction of oxidative DNA damage in isolated DNA and cultured rat hepatocytes. Mutat Res 518:123–133

Ling T, Fangke Y, Jun R (2010) Effect of mercury to seed germination, coleoptile growth and root elongation of four vegetables. Res J Phytochem 4:225–233

Liu X, Zhang S, Shan X, Zhu YG (2005) Toxicity of arsenate and arsenite on germination, seedling growth and amylolytic activity of wheat. Chemosphere 61:293–301

Llorens N, Arola L, Bladé C, Mas A (2000) Effects of copper exposure upon nitrogen metabolism in tissue cultured *Vitis vinifera*. Plant Sci 160:159–163

MacRobbie EA, Kurup S (2007) Signalling mechanisms in the regulation of vacuolar ion release in guard cells. New Phytol 175:630–640

Maksymiec W, Baszynski T (1999) The role of Ca^{2+} ions in modulating changes induced in bean plants by an excess of Cu^{2+} ions. Chlorophyll fluorescence measurements. Physiol Plant 105:562–568

Malecka A, Jarmuszkiewicz W, Tomaszewska B (2001) Antioxidative defense to lead stress in subcellular compartments of pea root cells. Acta Biochim Polon 48:687–698

Masclaux C, Valadier M, Brugière N, Morot-Gaudry J, Hirel B (2000) Characterization of the sink/source transition in tobacco (*Nicotiana tabacum* L.) shoots in relation to nitrogen management and leaf senescence. Planta 211:510–518

Meharg AA, Hartley-Whitaker J (2002) Arsenic uptake and metabolism in arsenic resistant and nonresistant plant species. New Phytol 154:29–43

Mishra S, Srivastava S, Dwivedi S, Tripathi RD (2011) Investigation of biochemical responses of Bacopa monnieri L. upon exposure to arsenate. Environ Toxicol. doi:10.1002/tox.20733

Molas J (1997) Changes in morphological and anatomical structure of cabbage (Brassica oleracera L.) outer leaves and in ultrastructure of their chloroplasts caused by an in vitro excess of nickel. Photosynthetica 34:513–522

Molas J (2002) Changes of chloroplast ultrastructure and total chlorophyll concentration in cabbage leaves caused by excess of organic Ni II complexes. Environ Exp Bot 47:115–126

Mysliwa-Kurdziel B, Prasad MNV, Strzalka K (2004) Photosynthesis in heavy metal stressed plants. In: Prasad MNV (ed) Heavy metal stress in plants: from biomolecules to ecosystems. Narosa Publishing House, New Delhi

Nriagu JO (1996) A history of global metal pollution. Science 272:223–224

Paivoke AEA, Simola LK (2001) Arsenate toxicity to Pisum sativum: mineral nutrients, chlorophyll content and phytase activity. Ecotox Environ Safe 49:111–121

Panda SK, Choudhury S (2005) Chromium stress in plants. Braz J Plant Physiol 17:95–102

Panda SK, Patra HK (2000) Nitrate and ammonium ions effect on the chromium toxicity in developing wheat seedlings. P Natl Acad Sci India B 70:75–80

Parys E, Romanowska E, Siedlecka M, Poskuta JW (1998) The effect of lead on photosynthesis and respiration in detached leaves and in mesophyll protoplasts of Pisum sativum. Acta Physiol Plant 20:313–322

Patra M, Bhowmik N, Bandopadhyay B, Sharma A (2004) Comparison of mercury, lead and arsenic with respect to genotoxic effects on plant systems and the development of genetic tolerance. Environ Exp Bot 52:199–223

Pätsikkä E, Kairavuo M, Sersen F, Aro E-M, Tyystjärvi E (2002) Excess copper predisposes photosystem II to photoinhibition in vivo by outcompeting iron and causing decrease in leaf chlorophyll. Plant Physiol 129:1359–1367

Peng XX, Yu XL, Li MQ, Yamauchi M (1996) Induction of proxidase by Fe2+ in detached rice leaves. Plant Soil 180:159–163

Perfus-Barbeoch L, Leonhardt N, Vavaddeur A, Forestier C (2002) Heavy metal toxicity: Cadmium permeates through calcium channels and disturbs the plant water status. Plant J 32:539–548

Pigna M, Cozzolino V, Violante A, Meharg AA (2009) Influence of phosphate on the arsenic uptake by wheat (*Triticum durum* L.) irrigated with arsenic solutions at three different concentrations. Water Air Soil Pollut 197:371–380

Pillay SV, Rao VS, Rao KVN (1996) Effect of nickel toxicity in *Hyptis suareeolens* (L.) Poit. and *Helianthus annuus* L. Indian J Plant Physiol 1:153–156

Ponnamperuma FN, Tianco EM, Loy T (1967) Redox equilibria in flooded soils: the iron hydroxide systems. Soil Sci 103:374–382

Prasad K, Saradhi PP, Sharmila P (1999) Concerted action of antioxidant enzymes and curtailed growth under zinc toxicity in *Brassica juncea*. Environ Exp Bot 42:1–10

Prasad MNV, Greger M, Landberg T (2001) *Acacia nilotica* L bark removes toxic elements from solution: corroboration from toxicity bioassay using *Salix viminalis* L in hydroponic system. Int J Phytoremed 3:289–300

Rai AN, Srivastava S, Paladi R, Suprasanna P (2011a) Calcium supplementation modulates arsenic-induced alterations and augments arsenic accumulation in callus cultures of Indian mustard (*Brassica juncea* (L.) Czern.). Protoplasma

Rai R, Pandey S, Rai SP (2011b) Arsenic-induced changes in morphological, physiological, and biochemical attributes and artemisinin biosynthesis in *Artemisia annua*, an antimalarial plant. Ecotoxicology 20:1900–1913

Rao KVM, Sresty TVS (2000) Antioxidative parameters in the seedlings of pigeonpea *(Cajanus cajan L.)* Millspauga in response to Zn and Ni stress. Plant Sci 157:113–128

Reeves RD, Baker AJM (2000) Metal-Accumulating Plants. In: Raskin I, Ensley BD (eds) Phytoremediation of toxic metals: using plants to clean up the environment. Wiley, New York

Ren F, Liu T, Liu H, Hu B (1993) Influence of zinc on the growth, distribution of elements, and metabolism of one-year old American ginseng plants. J Plant Nut 16:393–405

Romero-Puertas MC, Palma JM, Gomez LA, del Rio LA, Sandalio LM (2002) Cadmium causes oxidative modification of proteins in plants. Plant Cell Environ 25:677–686

Romero-Puertas MC, Rodriguez-Serrano M, Corpas FJ, Gomez M, del Rio LA, Sandalio LM (2004) Cadmium-induced subcellular accumulation of O_2^- and H_2O_2 in pea leaves. Plant Cell Environ 27:1122–1134

Roncel M, Ortega JM, Losada M (2001) Factors determining the special redox properties of photosynthetic cytochrome b559. Eur J Bioche 268:4961–4968

Sabat SC (1996) Copper iron inhibition of electron transport activity in sodium chloride washed photosystem II particle is partially prevented by calcium ion. Z Naturforsch 51c:179–184

Roth U, Von Roepenack-Lahaye E, Clemens S (2006) Proteome changes in Arabidopsis thaliana roots upon exposure to Cd^{2+}. J Exp Bot 57:4003–4013

Salt DE, Kato N, Kramer U, Smith RD, Raskin I (2000) The role of root exudates in nickel hyperaccumulation and tolerance in accumulator and nonaccumulator species of *Thlaspi*. In: Terry N, Banuelos G (eds) Phytoremediation of contaminated soil and water. CRS Press LLC, London

Samantaray S, Rout GR, Das P (2001) Induction, selection and characterization of Cr and Ni-tolerant cell lines of *Echinochloa colona* (L) in vitro. J Plant Physiol 158:1281–1290

Sanità di Toppi L, Gabbrielli R (1999) Response to cadmium in higher plants. Environ Exp Bot 41:105–130

Scandalios JG (1990) Response of plant antioxidant defense genes to environmental stress. Adv Genet 28:1–41

Schuster E (1991) The behavior of mercury in the soil with special emphasis on complexation and adsorption processes—A review of the literature. Water Air Soil Pollut 56:667–680

Schützendübel A, Polle A (2002) Plant responses to abiotic stresses: heavy metal-induced oxidative stress and protection by mycorrhization. J Exp Bot 53:1351–1365

Schützendübel A, Schwanz P, Teichmann T, Gross K, Langenfeld- Heyser R, Godbold DL et al (2001) Cadmium-induced changes in antioxidative systems, hydrogen peroxide content, and differentiation in Scots pine roots. Plant Physiol 75:887–898

Sersen F, Kralova K, Bumbalova A (1998) Action of mercury on the photosynthetic apparatus of spinach chloroplasts. Photosynthetica 35:551–559

Shanker AK, Pathmanabhan G (2004) Speciation dependant antioxidative response in roots and leaves of Sorghum (Sorghum bicolor (L) Moench cv CO 27) under Cr(III) and Cr(VI) stress. Plant Soil 265:141–151

Shanker AK, Cervantes C, Loza-Tavera H, Avudainayagam S (2005) Chromium toxicity in plants. Environ Int 31:739–753

Sharma CP (2006) Plant micronutrients. Science Publishers, Enfield

Sharma P, Dubey RS (2005) Lead toxicity in plants. Braz J Plant Physiology 17:35–52

Sharma DC, Sharma CP (1993) Chromium uptake and its effects on growth and biological yield of wheat. Cereal Res Commun 21:317–321

Sharma DC, Sharma CP (1996) Chromium uptake and toxicity effects on growth and metabolic activities in wheat, Triticum aestivum L. cv.UP Indian. J Exp Biol 34:689–691

Sharma Y, Leòn J, Raskin I, Davis KR (1996) Ozone-induced responses in Arabidopsis thaliana: the role of salicylic acid in the accumulation of defence-related transcripts and induced resistance. Proc Natl Acad Sci U S A 93:5099–5104

Sharma P, Bhardwaj R, Arora N, Arora HK, Kumar A (2008) Effects of 28-homobrassinolide on nickel uptake, protein content and antioxidative defence system in Brassica juncea. Biol Plant 52:767–770

Shi Q, Zhu Z (2008) Effects of exogenous salicylic acid on manganese toxicity, element contents and antioxidative system in cucumber. Environ Exp Bot 63:317–326

Singh AK (2001) Effect of trivalent and hexavalent chromium on (spinach Spinacea oleracea L). Environ Eco 119:807–810

Sinha S, Mukherji S, Dutta J (2002) Effect of manganese toxicity on pigment contet, Hill activity and photosynthetic rate of Vigna radiata L. Wilczek seedlings. J Envirn Biol 23:253–257

Smith SE, Christophersen HM, Pope S, Smith FA (2010) Arsenic uptake and toxicity in plants: integrating mycorrhizal influences. Plant Soil 327:1–21

Subrahmanyam D, Rathore VS (2001) Influence of Manganese toxicity on photosynthesis in ricebean (Vigna umbellata) seedlings. Photosynthetica 38:449–453

Suseela MR, Sinha S, Singh S, Saxena R (2002) Accumulation of chromium and scanning electron microscopic studies in scirpus lacustris l treated with metal and tannery effluent. Bull Environ Contam Toxicol 68:540–548

Teige M, Huchzermeyer B, Schultz G (1990) Inhibition of chloroplast ATPsynthase/ATPase is a primary effect of heavy metal toxicty in spinach plants. Biochem Physiol Pfl 186:165–168

Terry N, Zayed AM, de Souza MP, Tarun AS (2000) Selenium in higher plants. Annu Rev Plant Physiol Plant Mol Biol 51:401–432

Thordal-Christensen H, Zang Z, Wei Y, Collinge DB (1997) Subcellular localization of H_2O_2 accumulation in papillae and hypersensitive response during the barley powdery mildew interaction. Plant J 11:1187–1194

Trelease SF, Trelease HM (1939) Physiological differentiation in Astragalus with reference to selenium. Am J Bot 26:530–535

Tu C, Ma LQ (2005) Effects of arsenic on concentration and distribution of nutrients in the fronds of the arsenic hyperaccumulator Pteris vittata L. Environ Pollut 135:333–340

Van Assche F, Clijsters H (1986) Inhibition of photosynthesis in Phaseolus vulgaris by treatment with toxic concentrations of zinc: effects on electron transport and photo-phosphorylation. Physiol Plant 66:717–721

Verma S, Dubey RS (2003) Lead toxicity induces lipid peroxidation and alters the activity of antioxidant enzymes in growing rice plants. Plant Sci 164:645–655

Vromman D, Flores-Bavestrello A, Slejkovec Z, Lapaille S, Teixeira-Cardoso C, Briceño M, Kumar M, Martínez JP, Lutts S (2011) Arsenic accumulation and distribution in relation to young seedling growth in *Atriplex atacamensis* Phil. Sci Total Environ 412–413:286–295

Wang YD, Wang X, Wong YS (2012) Proteomics analysis reveals multiple regulatory mechanisms in response to selenium in rice. J Proteomics. http://dx.doi.org/10.1016/j.jprot.2011.12.030

Watanabe ME (1997) Phytoremediation on the brink of commercialization. Environ Sci Technol 31:182–186

Welch RM (1981) The biological significance of nickel. J Plant Nutr 3:345–356

Wu L, Huang ZZ (1992) Selenium assimilation and nutrient element uptake in white clover and tall fescue under the influence of sulphate concentration and Selenium tolerance of the plants. J Exp Bot 43:549–555

Yruela I (2005) Copper in plants. Braz J Plant Physiol 17:145–156

Yusuf M, Fariduddin Q, Hayat S, Ahmad A (2011) Nickel: an overview of uptake, essentiality and toxicity in plants. Bull Environ Contam Toxicol 86:1–17

Zeid IM (2001) Responses of Phaseolus vulgaris chromium and cobalt treatments. Biol Plant 44:111–115

Zhang WH, Tyerman SD (1999) Inhibition of water channels by $HgCl_2$ in intact wheat root cells. Plant Physiol 120:849–857

Zhang J, Zhao QZ, Duan GL, Huang YC (2011) Influence of sulphur on arsenic accumulation and metabolism in rice seedlings. Environ Exp Bot 72:34–40

Zhou SZ, Huang SQ, Guo K, Mehta SK, Zhang PC, Yang ZM (2007) Metabolic adaptations to mercury-induced oxidative stress in roots of *Medicago sativa* L. J Inorg Biochem 101:1–9

Zhou SZ, Wang SJ, Yang ZM (2008) Biological detection and analysis of mercury toxicity to alfalfa (*Medicago sativa*) plants. Chemosphere 70:1500–1509

Chapter 2
Plant Responses to Heavy Metal Toxicity

Anna Manara

Abstract Plants, like all other organisms, have evolved different mechanisms to maintain physiological concentrations of essential metal ions and to minimize exposure to non-essential heavy metals. Some mechanisms are ubiquitous because they are also required for general metal homeostasis, and they minimize the damage caused by high concentrations of heavy metals in plants by detoxification, thereby conferring tolerance to heavy metal stress. Other mechanisms target individual metal ions (indeed some plants have more than one mechanism to prevent the accumulation of specific metals) and these processes may involve the exclusion of particular metals from the intracellular environment or the sequestration of toxic ions within compartments to isolate them from sensitive cellular components. As a first line of defense, many plants exposed to toxic concentrations of metal ions attempt to prevent or reduce uptake into root cells by restricting metal ions to the apoplast, binding them to the cell wall or to cellular exudates, or by inhibiting long distance transport. If this fails, metals already in the cell are addressed using a range of storage and detoxification strategies, including metal transport, chelation, trafficking, and sequestration into the vacuole. When these options are exhausted, plants activate oxidative stress defense mechanisms and the synthesis of stress-related proteins and signaling molecules, such as heat shock proteins, hormones, and reactive oxygen species.

Keywords Metal homeostasis · Metal uptake · Metal transport · Chelation · Oxidative stress

A. Manara (✉)
Department of Biotechnology, University of Verona,
Strada Le Grazie 15, 37134 Verona, Italy
e-mail: anna.manara@univr.it

A. Furini (ed.), *Plants and Heavy Metals*, SpringerBriefs in Biometals,
DOI: 10.1007/978-94-007-4441-7_2, © Manara 2012

2.1 Signal Transduction in Response to Heavy Metals

The response to heavy metal stress involves a complicated signal transduction network that is activated by sensing the heavy metal, and is characterized by the synthesis of stress-related proteins and signaling molecules, and finally the transcriptional activation of specific metal-responsive genes to counteract the stress (Maksymiec 2007). The relevant signal transduction pathways include the Ca-calmodulin system, hormones, ROS signaling, and the mitogen-activated protein kinase (MAPK) phosphorylation cascade, which converge by activating the above-mentioned stress-related genes. Different signaling pathways may be used to respond to different heavy metals (DalCorso et al. 2010).

2.1.1 The Ca-Calmodulin System

Ca^{2+} signaling features in responses to a number of abiotic stress factors, including temperature extremes, osmotic stress, oxidative stress, anoxia, and mechanical perturbation (Knight 1999). Excess heavy metals modify the stability of Ca channels, thus increasing calcium flux into the cell. Intracellular Ca is a secondary messenger, which interacts with calmodulin to propagate the signal and ultimately to regulate downstream genes involved in heavy metal transport, metabolism, and tolerance (Yang and Poovaiah 2003). Higher intracellular Ca levels are observed in plants exposed to Cd, inducing adaptive mechanisms that alleviate the toxic effects of the heavy metal (Skórzyńska-Polit et al. 1998). The Ca-calmodulin system is also involved in the response to other heavy metal toxicity, such as Ni and Pb. Transgenic tobacco plants expressing NtCBP4 (*Nicotiana tabacum* calmodulin-binding protein) tolerate higher levels of Ni^{2+} but are hypersensitive to Pb^{2+}, reflecting the exclusion of Ni^{2+} but the accumulation of more Pb^{2+} than wild-type plants (Arazi et al. 1999).

2.1.2 Hormones in the Heavy Metal Response

Plant hormones are involved in many physiological and developmental processes, and play a crucial role in the adaptation to abiotic stress as shown by the regulation of hormone synthesis in the presence of heavy metals (Peleg and Blumwald 2011). For example, plants exposed to toxic levels of Cd, Cu, Fe, and Zn produce higher levels of ethylene, but Co does not have the same effect (Maksymiec 2007; Wise and Naylor 1988). Cd and Cu stimulate ethylene synthesis by upregulating ACC synthase expression and activity (Pell et al. 1997). Cu and Cd also induce the rapid accumulation of jasmonic acid (JA) in *Phaseolus coccineus* (Maksymiec et al. 2005), and Cu has also been shown to have this effect in rice (Rakwall et al. 1996)

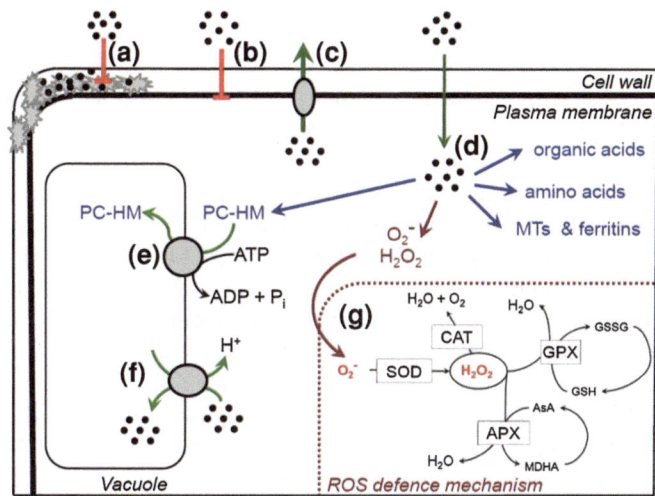

Fig. 2.1 The response to heavy metal toxicity in higher plants. As discussed in the text, plant responses to heavy metals include: **a** metal ion binding to the cell wall and root exudates; **b** reduction of metal influx across the plasma membrane; **c** membrane efflux pumping into the apoplast; **d** metal chelation in the cytosol by ligands such as phytochelatins, metallothioneins, organic acids, and amino acids; **e** transport of metal–ligand complexes through the tonoplast and accumulation in the vacuole; **f** sequestration in the vacuole by tonoplast transporters; **g** induction of ROS and *oxidative stress* defense mechanisms as described in Fig. 1.2. Metal ions are shown as *black dots*

and *Arabidopsis thaliana* (Maksymiec et al. 2005). Salicylic acid (SA) is involved in heavy metal stress responses, as shown by the increase in SA levels in barley roots in the presence of Cd and the ability of exogenous SA to protect roots from lipid peroxidation caused by Cd toxicity (Metwally et al. 2003).

2.1.3 The Role of Reactive Oxygen Species

As stated above, one of the major consequences of heavy metal accumulation is the production of ROS, which as well as causing widespread damage may also function as signaling molecules. Heavy metals such as Cd can produce ROS directly via the Fenton and Haber–Weiss reactions, and indirectly by inhibiting antioxidant enzymes (Romero-Puertas et al. 2007). In particular, H_2O_2 acts as a signaling molecule in response to heavy metals and other stresses (Dat et al. 2000). H_2O_2 levels increase in response to Cu and Cd treatment in *A. thaliana* (Maksymiec and Krupa 2006), upon Hg exposure in tomato (Cho and Park 2000) and in response to Mn toxicity in barley (Cho and Park 2000). This increase in H_2O_2 accumulation changes the redox status of the cell and induces the production of antioxidants and the activation of antioxidant mechanisms (Fig. 2.1).

2.1.4 The MAPK Cascade

The MAPK cascade in plants is a response to both biotic and abiotic stresses, including pathogens, temperature extremes, heavy metal stress, drought, and wounding (Jonak et al. 1996; He et al. 1999). The pathway is also used in hormone signal transduction and in response to developmental stimuli (Jonak et al. 2002). The MAPK cascade involved three kinases sequentially activated by phosphorylation: the MAPK kinase kinase (MAPKKK), the MAPK kinase (MAPKK), and the MAPK. At the end of this cascade of phosphorylation, MAPKs phosphorylate different substrates in different cellular compartments, including transcription factors in the nucleus. In this way, the MAPK cascade allows the transduction of the information to downstream targets. Four isoforms of MAPK were shown to be activated in alfalfa (*Medicago sativa*) seedlings exposed to Cu or Cd (Jonak et al. 2004) and a MAPK gene is also activated by Cd treatment in rice (Yeh et al. 2004). All these signaling pathways finally converge in the regulation of transcription factors that activate genes required for stress adaptation, particularly in the context of heavy metals this means genes for the activation of metal transporters and the biosynthesis of chelating compounds.

2.2 Metal Ion Uptake from Soil

Metal availability and motility in the rhizosphere is influenced by root exudates and microorganisms (Wenzel et al. 2003). Higher plants possess highly effective systems for the acquisition of metal ions and other inorganic nutrients from the soil. These are based on a small number of transport mechanisms, suggesting that different heavy metal cations are co-transported across the plasma membrane in the roots. Because toxic heavy metals such as Cd and Pb have no known biological function, it is likely that specific transporters do not exist. Instead, these toxic metals enter into the cells through cation transporters with a wide range of substrate specificity.

2.2.1 Metal Ion Binding to Extracellular Exudates and to the Cell Wall

As a first line of defense against heavy metals, plant roots secrete exudates into the soil matrix. One of the major roles of root exudates is to chelate metals and to prevent their uptake inside the cells (Marschner 1995). For example, Ni-chelating histidine and citrate are present in root exudates and these reduce the uptake of Ni from soil (Salt et al. 2000). The binding of metal ions such as Cu and Zn in the apoplast also helps to control the metal content of root cells (Dietz 1996). Cation binding sites are also present on the root cell wall, and this allows metal exchange thus influencing the availability of ions for uptake and diffusion into the apoplast (Allan and Jarrel 1989). The cell wall can play a key role in the immobilization of toxic heavy metal ions by providing pectic sites and hystidyl groups, and extra-cellular carbohydrates such as callose and mucilage, and thus prevents heavy metals uptake into the cytosol (Fig. 2.1). Thus, different tobacco genotypes with chemically distinct root cell wall surfaces have different sensitivities to Mn toxicity (Wang et al. 1992). These data suggest that the chemical properties of the cell wall might modulate plant metal uptake and consequently metal tolerance. However, the role of the cell wall in metal tolerance is not completely understood. The cell wall is in direct contact with metal ions in the soil but only a limited number of absorption sites are available, suggesting the cell wall has only a minor impact on metal tolerance (Ernst et al. 1992). However, *Silene vulgaris* ssp. *humilis* is a heavy metal-tolerant plant that accumulates different heavy metals by binding them to proteins or silicates in the epidermal cell walls (Bringezu et al. 1999).

2.2.2 Metal Ion Transport through the Plasma Membrane in Root

Plants possess various families of plasma membrane transporters involved in metal uptake and homeostasis. At the cellular level, metal transporters on the plasma membrane and tonoplast are required to maintain physiological concentrations of heavy metals, but they may also contribute to heavy metal stress responses (Fig. 2.1). These transporters belong to the heavy metal P_{1B}-ATPase, the NRAMP, the CDF (Williams et al. 2000), and the ZIP families (Guerinot 2000). The biological function, cellular location, and metal specificity of most of these transporters in plants are still unknown. In plants most of these metal ions transporters were identified by complementation in *Saccharomyces cerevisiae* mutants defective in metal uptake.

2.2.2.1 The ZIP Family

One of the principal metal transporter family involved in metal uptake is the ZIP family. ZIP family of transporters have been identified in many plant species (as well as bacteria, fungi and animals) and are involved in the translocation of divalent cations across membranes. Certain ZIP proteins are induced in *A. thaliana* roots and shoots in response to Fe or Zn loading, and thus appear to be part of a stress response. Most ZIP proteins are predicted to comprise eight transmembrane domains and have a similar topology, with the N- and C-termini exposed to the apoplast, and a variable cytoplasmic loop between transmembrane domains III and IV that contains a histidine-rich domain putatively involved in metal binding (Guerinot 2000) and specificity (Nishida et al. 2008).

The first ZIP transporter to be characterized was the *A. thaliana* IRT1. This was identified by functional complementation of the *S. cerevisiae fet3fet4* double mutant, which is impaired in iron transport (Eide et al. 1996). In *A. thaliana*, IRT1 is expressed in root cells and accumulates in response to iron deficiency, suggesting a role in Fe^{2+} uptake from the soil (Vert et al. 2002). Many metal transporters present low ion selectivity, and additional studies in yeast showed that AtIRT1 can also transport Mn^{2+}, Zn^{2+}, and Cd^{2+} (Korshunova et al. 1999). IRT1, in plant, is also involved in the uptake of heavy metal divalent cations such as Cd^{2+} and Zn^{2+} (Cohen et al. 1998). Furthermore, when expressed in yeast, *AtIRT1* enhanced the Ni^{2+}-uptake activity (Nishida et al. 2011). In *A. thaliana*, *AtIRT1* is induced in response to excess Ni and is involved in Ni^{2+} transport and accumulation. In *S. cerevisiae* the ZRT1 and ZRT2 transporters were identified on the basis of sequence similarity to IRT1 and they are respectively high- and low-affinity Zn^{2+} transporters (Zhao and Eide 1996a, b). The *zrt1zrt2* double mutant yeast was then used to clone the *A. thaliana* Zn^{2+} transporters, AtZIP1, AtZIP2, and AtZIP3 by functional complementation (Grotz et al. 1998). *ZIP1* and *ZIP3* are expressed principally in the roots and are induced under Zn limiting conditions. In *A. thaliana*, the analysis of the genomic sequence revealed the presence of a fourth member of the family, *AtZIP4*, which is expressed in roots and shoots, and it is also induced by Zn restriction, supporting their proposed role in Zn nutrition. ZIP transporters in plants are also involved in Cd uptake from soil into the root cells and transport Cd from root to shoot (Krämer et al. 2007). In hyperaccumulator species, ZIP transporters are necessary (but not sufficient) for the enhanced accumulation of metal ions and metal accumulating capacity correlates with ZIP expression (Krämer et al. 2007). In *S. cerevisiae*, ZRT3 is a further transporter identified by functional complementation but this appears to be involved in the mobilization of Zn from vacuole and not only in the uptake from the environment (MacDiarmid et al. 2000). There is some evidence that Ni is taken up by Zn transporters (Assunção et al. 2001) although candidate Ni-specific transporters have also been identified (Peer et al. 2003).

2.2.2.2 The NRAMP Family

NRAMP metal transporters have been shown to transport a wide range of metals, such as Mn^{2+}, Zn^{2+}, Cu^{2+}, Fe^{2+}, Cd^{2+}, Ni^{2+}, and Co^{2+}, across membranes, and have been identified in bacteria, fungi, plants, and animals (Nevo and Nelson 2006). In plants, NRAMP transporters are expressed in roots and shoots and are involved in transport of metal ions through the plasma membrane and the tonoplast (Krämer et al. 2007). NRAMPs in *A. thaliana* are thought to transport Fe and Cd, with NRAMP1 playing a specific role in Fe transport and homeostasis (Thomine et al. 2000). The *AtNramp1* gene complements the yeast *fet3fet4* double mutant, and is induced under limiting Fe conditions. (Curie et al. 2000). The overexpression of *AtNramp1* in transgenic *A. thaliana* plants leads to an increase in plant resistance to toxic iron concentration (Curie et al. 2000).

2.2.2.3 The Copper Transporters Family

The Copper Transporters (CTR) family of transporter has firstly been identified in yeast and mammalian and subsequently also in plants. CTR proteins comprise a putative metal-binding motif in the extracellular domain, three predicted transmembrane domains, and a conserved and essential MXXXM motif within the putative second transmembrane domain (Puig and Thiele 2002). The *A. thaliana* copper transporter COPT1 was identified by functional complementation of the *S. cerevisiae* mutant *ctr1-3*, which is defective in copper uptake (Kampfenkel et al. 1995a). In *A. thaliana*, COPT1 has been shown to transport copper, and it also has a role in growth and pollen development (Sancenón et al. 2004).

2.2.3 Reduced Metal Uptake and Efflux Pumping at the Plasma Membrane

The plasma membrane plays an important role in plant response to heavy metals by preventing or reducing the uptake of metals into the cell or by active efflux pumping outside the cell. There are few examples of ion exclusion or reduced uptake as a sole protective mechanism in plants. Although an arsenate-tolerant genotype of *Holcus lanatus* absorbs less arsenate than an equivalent non-tolerant genotype (Meharg and Macnair 1992) due to the suppression of the high-affinity arsenate transport system combined with the constitutive synthesis of PCs (Hartley-Whitaker et al. 2001). Active efflux systems are more common and are used to control heavy metals accumulation inside the cell. This mechanism is well documented in bacteria (Silver 1996) and in animal cells (Palmiter and Findley 1995). Differently, there are only few evident indications of plasma membrane efflux transporters involved in heavy metal response in plants. Comparing data obtained for bacteria and mammals, the most likely candidate heavy metal efflux pumps in plants (based on sequence similarity to microbial and animal proteins) are the P_{1B}-ATPases and the CDF families of transporters. P_{1B}-type ATPases belong to P-type ATPase superfamily and use energy from ATP hydrolysis to translocate diverse metal cations across biological membranes (Axelsen and Palmgren 2001). P_{1B}-type ATPases share common structural characteristics, such as eight predicted transmembrane domains, a CPx (Cys-Pro-Cys/His/Ser) intramembrane motif that is hypothetically involved in metal translocation (Ashrafi et al. 2011), and a putative N- or C-terminal metal binding domain (Colangelo and Guerinot 2006). P_{1B}-ATPases pump metal ions out of the cytoplasm against their electrochemical gradient, into either the apoplast or into the vacuole. The eight P_{1B}-type ATPases in *A. thaliana* and rice were renamed heavy metal ATPases (HMAs) (Baxter et al. 2003). HMAs are divided into two classes, one involved in transport of monovalent cations (Cu/Ag) and the second in the transport of divalent cations (Zn/Co/Cd/Pb) (Baxter et al. 2003). HMAs are more selective than the transporters involved in metal uptake, e.g., HMA2, HMA3, and HMA4 export Zn and Cd exclusively (Krämer et al. 2007). Therefore, *hma2 hma4* double mutants and, to a lesser extent, the *hma4* single mutant contain low levels of Zn in the shoots, display Zn deficiency symptoms, but other micronutrients are unaffected (Hussain et al. 2004). In addition, they show increased Cd sensitivity and decrease in Cd root-to-shoot translocation (Hussain et al. 2004; Wong and Cobbett 2009). Both AtHMA2 (Hussain et al. 2004) and AtHMA4 (Verret et al. 2005) are located on the plasma membrane and heterologous expression of AtHMA4 in yeast induces tolerance to Zn and Cd toxicity, thus suggesting that this transporter can act as efflux pump (Mills et al. 2005).

ABC transporters are also involved in metal ion efflux from the plasma membrane. For example, AtPDR8 is localized in the plasma membrane of *A. thaliana* root hairs and epidermal cells, conferring both metal tolerance (Kim et al. 2007) and pathogen resistance (Kobae et al. 2006). AtPDR8 is induced in the presence of Cd and Pb, transgenic plants overexpressing the protein do not accumulate Cd in the roots or shoots and are tolerant to normally toxic levels of Cd and Pb. In contrast, mutants accumulate higher levels of Cd and are sensitive to both metals. Probably, AtPDR8 acts as an efflux pump of these metals at the plasma membrane (Kim et al. 2007).

2.3 Root-to-Shoot Metal Translocation

Once taken up by the roots, metal ions are loaded into the xylem and transported to the shoots as complexes with various chelators. Organic acids, especially citrate, are the major chelators for Fe and Ni in the xylem (Tiffin 1970; Leea et al. 1977). In addition, amino acids are potential metal ligands, for instance Ni may also be chelated by histidine and translocated (Kramer et al. 1996), and the methionine derivative NA is involved in the transport of Cu (Pich and Scholz 1996). Several types of transporter proteins are involved in the root-to-shoot transport of metals. Metal ions are also translocated from source to sink tissue via phloem. Therefore, phloem sap contains metals arising from source tissue, like Fe, Cu, Zn, and Mn (Stephan et al. 1994). Into the phloem, only NA was identified as potential metal chelator of Fe, Cu, Zn, and Mn (Stephan and Scholz 1993). NA is involved in the long distance transport of metals inside the xylem and phloem, but other chelators are required for loading. High molecular weight compounds that chelate Ni, Co, and Fe are found in the phloem of *Ricinus communis* plants (Wiersma and Van Goor 1979; Maas et al. 1988) and Zn-chelating peptides are found in *Citrus* spp. (Taylor et al. 1988) but they have yet to be characterized in detail.

2.3.1 The HMA Family of Transporters

The P-type ATPases reclassified as HMAs (see above) function not only as efflux pumps to remove metal ions from the cell, but also as internal transporters to load Cd and Zn metals into the xylem from the surrounding tissues. HMA4 is the first gene encoding for P-type ATPase cloned and characterized in *A. thaliana* (Mills et al. 2003). AtHMA4 is a plasma membrane transporter of divalent ions required for Zn homeostasis and Cd detoxification as it participates in the cytosolic efflux and in the root-to-shoot translocation of these metals (Mills et al. 2003; Verret et al. 2004). Overexpression of the AtHMA4 protein not only increases Zn and Cd tolerance, but also enhances the root-to-shoot translocation of both metals suggesting a role also in metal root-to-shoot transport (Verret et al. 2004). AtHMA5 is expressed constitutively in roots and induced by Cu in other plant organs, but *hma5* mutants are hypersensitive to Cu and accumulate this metal in roots to a greater extent than wild-type plants, suggesting a role in root-to-shoot translocation and Cu detoxification (Andrés-Colás et al. 2006).

2.3.2 The MATE Family of Efflux Proteins

MATE is a family of membrane-localized efflux proteins involved in extrusion of multidrug and toxic compound from the cell. FRD3 is a MATE protein that participates in iron-citrate efflux, i.e., the loading of Fe^{2+} and citrate into the vascular tissue in the roots. Xylem exudates from *frd3* mutant plants contain less citrate and Fe than wild-type plants, whereas those from transgenic plants over-expressing FRD3 produce more citrate in root exudates. Ferric-citrate complexes are required for the translocation of Fe to the leaves because Fe moves through the xylem in its chelated form (Durrett et al. 2007).

2.3.3 The Oligopeptide Transporters Family

Oligopeptide Transporters (OPT) is a superfamily of oligopeptide transporters including the YSL subfamily. The YSL family, specific for plants, takes its name from the maize Yellow stripe 1 protein (ZmYS1) that mediates Fe uptake by transporting Fe(III)-phytosiderophore complexes (Curie et al. 2001). The ZmYS1 transporter is able to translocate Fe, Zn, Cu, Ni, and, to a lesser extent, Mn and Cd (Schaaf et al. 2004). These ions can be chelated by either phytosiderophores or NA (Roberts et al. 2004). Eight putative YSL transporters have been identified in *A. thaliana* based on similarity to the maize gene (Colangelo and Guerinot 2006). AtYSL1 is expressed in the leaf xylem parenchyma, in pollen, and young siliques; mutants have a low Fe-NA complex content and cannot germinate normally, indicating a role of this protein in the transport of chelated Fe (Le Jean et al. 2005). AtYSL2 is expressed in shoot and root vascular tissues and is localized mainly in the lateral plasma membrane, consistent with a role in the lateral movement of metals into the veins (DiDonato et al. 2004; Schaaf et al. 2005). AtYSL2 is modulated by Fe, Cu (DiDonato et al. 2004), and Zn (Schaaf et al. 2005) and can transport both Fe and Cu as NA complexes (DiDonato et al. 2004).

2.4 Heavy Metal Chelation in the Cytosol

Inside the cell, heavy metal ions that are not immediately required metabolically may reach toxic concentrations, and plant cells have evolved various mechanisms to store excess metals to prevent their participation in unwanted toxic reactions. If the toxic metal concentration exceeds a certain threshold inside the cells, an active metabolic process contributes to the production of chelating compounds. Specific peptides such as PCs and MTs are used to chelate metals in the cytosol and to sequester them in specific subcellular compartments. A large number of small molecules are also involved in metal chelation inside the cells, including organic acids, amino acids, and phosphate derivatives (Rauser 1999) (Fig. 2.1).

2.4.1 Phytochelatins

PCs are the best-characterized heavy metal chelators in plants, especially in the context of Cd tolerance (Cobbett 2000). PCs are a family of metal-binding peptides with the general structure $(\gamma\text{-Glu-Cys})_n\text{Gly}$ (n = 2–11) (Cobbett and Goldsbrough 2002). They are present in plants and fungi. The cysteine thiol groups allow PCs to chelate metals and form complexes with a molecular weight of 2.5–3.6 kDa (Cobbett 2000). PCs are synthesized in the cytosol and then transported as complexes to the vacuole. Their synthesis is rapidly activated in the presence of heavy metals such as Cd, Cu, Zn, Ag, Au, Hg, and Pb (Rauser 1995; Cobbett 2000). Synthesis involves the chain extension of GS by PCS a constitutively expressed cytosolic enzyme whose activity is controlled post-translationally because the metal ions chelated by PCs are required for enzyme activity (Grill et al. 1989; Cobbett 2000). Due to their metal affinity, PCs are thought to be involved in the homeostasis and trafficking of essential metals such as Cu and Zn (Thumann et al. 1991) and in the detoxification of heavy metals, but they do not seem to be involved in hyperaccumulation (Ebbs et al. 2002). Contrasting evidence has been reported for the role of PCs in heavy metal tolerance. They have a clear role in the response of plants and yeast to Cd, e.g., they are induced rapidly in *Brassica juncea* following the intracellular accumulation of Cd, thus protecting the photosynthetic apparatus despite a decline in transpiration and leaf expansion (Haag-Kerwer et al. 1999). Furthermore, the Cd sensitivity of various *A. thaliana* mutants correlated with their ability to accumulate PCs (Howden et al. 1995). Cd and Cu treatment also induces the transcription of genes involved in the synthesis of GS, the precursor of PCs (Xiang and Oliver 1998). Transgenic *A. thaliana* plants with low GS levels are more sensitive to Cd, whereas those with elevated GS levels have similar Cd tolerance to wild-type plants (Xiang et al. 2001). Similarly, overexpression of the *Escherichia coli* γ-glutamylcysteine synthetase gene in *B. juncea* increased the synthesis of GS and PCs, resulting in greater Cd tolerance (Zhu et al. 1999).

The *A. thaliana* gene for PC synthase (*CAD1*) was identified by using the Cd-sensitive, PC-deficient *cad1* mutant (Ha et al. 1999). *cad1* mutant produces normal levels of GSH but is deficient in PCs and hypersensitive to Cd (Howden et al. 1995). A *Schizosaccharomyces pombe* mutant with the same characteristics has also been identified (Ha et al. 1999). The expression of *AtPCS1* from *A. thaliana* and *TaPCS1* from wheat in *S. cerevisiae* increases PC synthesis and induces Cd tolerance (Vatamaniuk et al. 1999; Clemens et al. 1999). Furthermore, purified recombinant *A. thaliana* and *S. pombe* PC synthases catalyze the production of PCs from GSH in vitro (Vatamaniuk et al. 1999; Clemens et al. 1999).

To evaluate the role of PCs in the heavy metal stress response, the sensitivity of *cad1* mutants to different metals was analyzed. The *A. thaliana cad1–3* mutant is more sensitive to arsenate and Cd than wild-type plants, while there is no difference considering Zn, selenite, and Ni ions (Ha et al. 1999). In contrast to the *S. pombe pcs* mutant, *cad1–3* was also slightly sensitive to Cu and Hg and showed intermediate sensitivity to Ag (Ha et al. 1999). The role of PCs in Cu tolerance remains to be determined. Studies of the copper-tolerant plant *Mimulus guttatus* confirmed a role for PCs in Cu tolerance (Salt et al. 1989). In contrast, Cu-sensitive and Cu-tolerant ecotypes of *S. vulgaris* produced similar amounts of PC when the root tips were exposed to Cu, suggesting differential tolerance arises from other mechanism (Schat and Kalff 1992; De Knecht et al. 1994). Therefore, although many evidences for the role of PCs in plant response and detoxification are reported, not all studies have supported an effective role of PCs in metal tolerance.

It is also notable that excessive PC levels in transgenic plants increase the accumulation of heavy metals without enhancing tolerance (Pomponi et al. 2006) and can even confer hypersensitivity to heavy metals. Indeed, an excessive expression of *AtPCS* genes confers a hypersensitivity to Cd stress (Lee et al. 2003). This probably reflects additional important roles for PCs in the cell, such as essential metal homeostasis, antioxidant mechanisms, and sulfur metabolism (Rauser 1995; Dietz et al. 1999; Cobbett 2000). Their role in heavy metal stress response probably may be a side effect of these functions (Steffens 1990).

The final stage of PC activity, particularly in the Cd response, involves their accumulation as complexes in the vacuole (Salt et al. 1998), where they eventually form HMW complexes after incorporation of S^{2-}. PC-Cd complexes are transported into the vacuole by Cd/H^+ antiporters and ATP-dependent ABC transporters in the tonoplast (Salt and Wagner 1993; Salt and Rauser 1995). In *S. pombe*, a Cd-sensitive mutant has been isolated; this strain can synthesize PCs but not accumulate the Cd-PC-sulfide complexes (Ortiz et al. 1992).

The mutant results to have a mutation in the gene *hmt1* that encodes for an ABC-type transporter, suggesting that this gene mediated transport and compartmentalization of heavy metals. Similar ABC-type transporters may also be involved in the compartmentalization of Cd in higher plants (Salt and Rauser 1995).

PCs also mediate the root-to-shoot transport of Cd. Transgenic *A. thaliana* plants expressing wheat *TaPCS1* accumulate small amounts of Cd in the roots but have an increased Cd transport to the shoot, reflecting the increased transport efficiency (Gong et al. 2003).

2.4.2 Metallothioneins and Ferritins

Like PCs, MTs are a major family of LMW cysteine-rich metal-binding peptides. MTs have been found in many organisms, although the MTs in plants differ considerably from those found in mammals and fungi. As they contain mercaptide groups they are able to bind metal ions. *Class 1* MTs contain cysteine motifs that align with mammalian MTs, whereas *Class 2* MTs contain similar cysteine clusters but they do not align with *Class1* MTs (Robinson et al. 1993). *Class 1* MTs are characterized by the exclusive presence of Cys–X–Cys motifs, whereas in *Class 2* MTs both Cys–Cys and Cys–XX–Cys pairs are located in the N-terminal domain (Robinson et al. 1993). In vertebrates, MTs contain a stretch of 20 highly conserved cysteine residues, whereas plant and fungal MTs do not contain this motif (Cherian and Chan 1993).

In *S. cerevisiae* the MT-encoding gene *CUP1* is synthesized and activated by metal ions, such as Cu (Fürst et al. 1988). In plants, MTs are induced by various abiotic stresses but are also expressed during development (Rauser 1999). In wheat and in rice, MTs are induced by metal ions, such as Cu and Cd, and by abiotic stresses such as temperature extremes and nutrient deficiency (Cobbett and Goldsbrough 2002). Plant MTs sequester excess of metals by coordinating metal ions with the multiple cysteine thiol groups (Robinson et al. 1993), and have particular affinity for Zn^{2+}, Cu^+, and Cu^{2+} as shown by the expression of the pea gene *PsMTa* in *E. coli* (Tommey et al. 1991). *A. thaliana* MT gene expression is activated in response to Cu and Cd treatment, but not by Zn, e.g., MT2 is expressed in response to Cu, but only marginally in the presence of Cd and Zn (Zhou and Goldsbrough 1994). In *A. thaliana,* MT1a and MT2a are expressed in the trichomes and the phloem, indicating they take part in both heavy metal sequestration and in metal ion transport (García-Hernández et al. 1998). *A. thaliana mt1a* mutants are hypersensitive to Cd and accumulate much lower levels of As, Cd, and Zn than wild-type plants, showing that MTs play a role in both metal tolerance and metal accumulation (Zimeri et al. 2005). This is supported by overexpression experiments, e.g., *Vicia faba* guard cells overexpressing *A. thaliana AtMT2a* and *AtMT3* can tolerate higher levels of Cd than untransformed cells (Lee et al. 2004). Similarly, the overexpression of *CcMT2* from legume *Cajanus cajan* in *A. thaliana* induces both Cd and Cu tolerance and allows both metals to accumulate without affecting the expression of endogenous transporters (Sekhar et al. 2011).

Although animal and fungal MTs have a clear role in heavy metal detoxification (Hamer 1986) the precise relationship between plant MTs and heavy metals is unknown (Zhou and Goldsbrough 1994; Zenk 1996; Giritch et al. 1998). MTs are known to participate in Cu homeostasis (Cobbett and Goldsbrough 2002) and *A. thaliana MT1* and *MT2* complement the *S. cerevisiae* MT-defective *cup1* mutant and confer Cd tolerance (Zhou and Goldsbrough 1994). *A. thaliana MT2* can also partially rescue the Zn hypersensitivity of a *Synechococcus amtA* mutant, which is deficient for an endogenous Zn^{2+}-MT gene (Robinson et al. 1996). The expression pea MT type I as a GSH transferase fusion in *E. coli* increases Cu accumulation (Evans et al. 1992), and the expression of *B. juncea* MT2 in *A. thaliana* enhances Cd and Cu tolerance (Zhigang et al. 2006). In contrast, MT2 expression is delayed in *B. juncea* plants exposed to Cd although there is a rapid induction of PC biosynthesis (Haag-Kerwer et al. 1999). These results indicate that there is no correlation between MT2 expression and Cd accumulation in leaves and the precise role of MTs in plants under heavy metal stress remains to be established. MTs probably have different functions in response to different heavy metals and could also participate in additional antioxidant protection mechanism and plasma membrane repair (Hamer 1986).

Ferritins are ubiquitous multimeric proteins that can store up to 4500 iron atoms in a central cavity (Harrison and Arosio 1996). Animal ferritins can also store other metals, including Cu, Zn, Cd, Be, and Al (Price and Joshi 1982; Dedman et al. 1992) whereas plant ferritins have only been shown to store Fe. Plant ferritins are synthesized in responses to various environmental stresses, including photoinhibition and iron overloading (Murgia et al. 2001, 2002). Ferritin gene expression in plants is dually regulated by ABA and by antioxidants and serine/threonine phosphatase inhibitors (Savino et al. 1997). Ferritins are therefore a front-line defense mechanism against free iron-induced oxidative stress (Ravet et al. 2009). The major function of plant ferritins is not to store and release iron, as previously reported, but to scavage free reactive iron and prevent oxidative damage (Ravet et al. 2009).

2.4.3 Organic Acids, Amino Acids, and Phosphate Derivatives

Organic acids and amino acids can bind heavy metals and may therefore be deployed in response to metal toxicity (Rauser 1999). However, a clear correlation between heavy metal accumulation and the production of these compounds has not been established. Organic acids such as malate, citrate, and oxalate confer metal tolerance by transporting metals through the xylem and sequestrating ions in the vacuole, but they have multiple additional roles in the cell (Rauser 1999).

Citrate, which is synthesized in plants by the enzyme citrate synthase, has a higher capacity for metal ions than malate and oxalate, and although its principal role is to chelate Fe^{2+} it also has a strong affinity also for Ni^{2+} and Cd^{2+} (Cataldo et al. 1988). Malate is a cytosolic Zn-chelator in zinc-tolerant plants (Mathys 1977).

Also amino acids and derivatives are able to chelate metals conferring to plants resistance to toxic levels of metal ions. Histidine is considered the most important free amino acid in heavy metal metabolism. Thanks to the presence of carboxyl, amino, and imidazole groups, it is a versatile metal chelator, which can confer Ni tolerance and enhance Ni transport in plants when supplied in the growth medium, perhaps reflecting its normal role as a chelator in root exudates (Callahan et al. 2006). Histidine levels also increase in the xylem of *Alyssum lesbiacum*, a Ni hyperaccumulator, when the plant is exposed to Ni (Kramer et al. 1996).

The amino acid derivative NA is an aminocarboxylate synthesized by the condensation of three S-adenosyl-L-methionine molecules in a reaction catalyzed by NA synthase (Shojima et al. 1990). NA chelates Fe, Cu, and Zn in complexes (Stephan et al. 1996) and then accumulates within vacuoles (Pich et al. 1997); it is not secreted from the roots (Stephan and Scholz 1993). NA is also involved in the movement of micronutrients in plants (Stephan and Scholz 1993). The physiological role of NA has been studied extensively in the tomato mutant *chloronerva*, which lacks a functional NA synthase gene and is characterized by the abnormal distribution and accumulation of Fe (Scholz et al. 1985) and Cu (Herbik et al. 1996). NA is also the precursor of the phytosiderophore mugineic acid, which binds Zn^{2+}, Cu^{2+}, and Fe^{3+} (Treeby et al. 1989). This derivative is synthesized in grasses by the deamination, reduction, and hydroxylation of NA (Shojima et al. 1990).

Phytate (myo-inositol hexakisphosphate) is the principal form of reserve phosphorous in plants (Hocking and Pate 1977) and is often localized in the roots and seeds (van Steveninck et al. 1993; Hubel and Beck 1996). The molecule comprises six phosphate groups which allow the chelation of multiple cations, including Ca^{2+}, Mg^{2+}, and K^+, but also Fe^{2+}, Zn^{2+} and Mn^{2+} (Mikus et al. 1992). The distribution of phytate and its ability to chelate multiple metal species suggest it could be mobilized as a detoxification strategy. In support of this, the addition of Zn to the culture medium leads to the production of Zn^{2+}-containing phytate globoids in the root endoderm and pericycle cells of certain crops (van Steveninck et al. 1993). Therefore, a controlled synthesis or mobilization of phytate in these cell layers plays a key role in metal ion loading to the aerial parts of plants.

2.5 Metal Sequestration in the Vacuole by Tonoplast Transporters

When metal ions are accumulated in excess inside the cytosol, plants have to remove them in order to minimize their toxic effects. Plants respond to high intracellular concentrations of metal ions by using efflux pumps either to export the ions to the apoplast (as discussed above) or to compartmentalize them within the cell. The main storage compartment for metal ions is the vacuole, which in plants accounts for up to 90% of the cell volume (Vögeli-Lange and Wagner 1990). Several families of intracellular transporters involved in this process have been identified in plants and yeast and they appear to be highly selective.

2.5.1 The ABC Transporters

ABC transporters can transport xenobiotics and heavy metals into the vacuole, and two subfamilies (MRP and PDR) are particularly active in the sequestration of chelated heavy metals. Plant cell vacuoles are, in fact, the major site for accumulation and storage of PC-Cd complexes. PC-Cd complexes are generated in the cytosol and are then translocated by ABC transporters (Vögeli-Lange and Wagner 1990). In the vacuoles more Cd and sulfide are incorporated to form HMW complexes, the main Cd storage form. The first vacuolar ABC transporter HMT1 was identified by its ability to complement a S. pombe mutant that cannot produce HMW complexes (Ortiz et al. 1992). HMT1 is localized in the tonoplast and transports PC-Cd complexes into the vacuole in a Mg-ATP-dependent manner (Ortiz et al. 1995). A similar protein has been identified in oat roots, but HMT1 homologs are yet to be found in other plants (Salt and Rauser 1995). In S. cereviasiae, the tonoplast ABC pump YCF1 transports Cd into the vacuole as a bis(glutathionato)Cd complex, and confers Cd tolerance (Li et al. 1997). MRP-related sequences like YCF1 have been found in A. thaliana (Lu et al. 1997, 1998) and are the most likely candidates for PC-Cd transport across the tonoplast because HMT1 homologs are scarce in plants. In A. thaliana, two transporters belonging to the ABC family, AtMRP1 and AtMRP2 have been shown to transport PC-Cd complexes into the vacuole but the role of AtMRP3 in Cd transport remains unclear (Lu et al. 1997, 1998). AtMRP3 partially complements yeast Δycf1 mutants but there is no evidence of PC-Cd transport into the vacuole (Tommasini et al. 1998). Moreover, many plants produce Mg-ATP-dependent transporters of GS-S-conjugates (Martinoia et al. 1993).

2.5.2 The CDF Transporters

Members of the CDF transporter family (also called MTPs in plants) are involved in the transport of metal ions from the cytoplasm to the vacuole (Krämer et al. 2007), and to the apoplast and endoplasmic reticulum (Peiter et al. 2007). CDF transporters have been characterized primarily in prokaryotes (Nies 1992) but are also found in many eukaryotes, where they transport divalent metal cations such as Zn, Cd, Co, Fe, Ni, and Mn (Montanini et al. 2007). Eukaryotic CDF transporters are characterized by six transmembrane domains, a C-terminal cation efflux domain, and a histidine-rich region between transmembrane domains IV and V (Mäser et al. 2001) that may act as a sensor of metal concentration (Kawachi et al.

2008). The CDF family can be divided into four phylogenetic groups (Mäser et al. 2001), but groups I and III are the most interesting in plants since these are the ones involved in metal tolerance and accumulation (Krämer et al. 2007). The *A. thaliana* ZAT1 transporter (later renamed AtMTP1) is closely related to the animal *ZnT* Zn transporter and its function is the vacuolar sequestration of Zn (van der Zaal et al. 1999). The gene is constitutively expressed and is not induced by Zn, but its overexpression in transgenic plants exposed to high levels of Zn confers resistance to Zn toxicity and leads to Zn accumulation in the roots without altering Cd sensitivity (van der Zaal et al. 1999). These data suggest that AtMTP1 transports Zn into the vacuole and may represent a Zn tolerance mechanism. Another tonoplast transporter, AtMTP3, is also involved in the transport of Zn into the vacuole (Kramer et al. 2007).

2.5.3 The HMA Transporters

As stated above, P_{1B}-ATPases (HMAs) are involved in the efflux of metal ions from the cytoplasm, and those on the tonoplast (such as AtHMA3) are thought to contribute to Cd and Zn homeostasis by sequestration into the vacuole (Krämer et al. 2007). However, AtHMA3 may play a role in the detoxification of a wider range of heavy metals through storage in the vacuoles, because overexpression induces tolerance to Cd, Pb, Co, and Zn (Morel et al. 2009).

2.5.4 CaCA Transporters

The CaCA superfamily is ubiquitous in both prokaryotes and eukaryotes and is an integral component of Ca^{2+} cycling systems that involve the efflux of Ca^{2+} across membranes against a concentration gradient. This is achieved by using a counter-electrochemical gradient of other ions, such as H^+, Na^+, or K^+ (Emery et al. 2012). Examples of members of the CaCA families that may be involved in metal homeostasis are MHX and CAX. MHX is a vacuolar Mg^{2+} and Zn^{2+}/H^+ antiport expressed predominantly in xylem-associated cells; overexpression in tobacco increases sensitivity to Mg and Zn although the concentration of these metals in shoots is unchanged (Shaul et al. 1999). The CAX family are Ca^{2+}/H^+ antiports that also recognize Cd^{2+}, suggesting this is an important route for Cd sequestration in the vacuole (Salt and Wagner 1993). In *A. thaliana*, only CAX proteins such as AtCAX2 and AtCAX4 seem to be involved in the vacuolar accumulation of Cd. The overexpression of *AtCAX2* and *AtCAX4* results in the accumulation of more Cd in the vacuoles (Korenkov et al. 2007). AtCAX4 is expressed mainly in root tips and primordia and is induced by Ni and Mn. Root growth in response to Cd and Mn is altered in *cax* mutants whereas overexpression induces symptoms that are compatible with Cd accumulation (Mei et al. 2009).

2.5.5 NRAMP Transporters

NRAMP transporters such as AtNRAMP3 and AtNRAMP4 are localized in the tonoplast and are probably functionally redundant. The *nramp3 nramp4* double mutant has a Fe-deficient phenotype in seedlings that can be rescued by providing excess Fe, although the Fe content is the same as in wild-type plants suggesting that AtNRAMP3 and AtNRAMP4 are required to mobilize Fe from the vacuole (Thomine et al. 2003; Lanquar et al. 2005). The overexpression of AtNRAMP3 increases Cd sensitivity (Thomine et al. 2000) and reduces the accumulation of Mn (Thomine et al. 2003), indicating a possible role in the homeostasis of metals other than Fe.

2.6 Oxidative Stress Defence Mechanisms and the Repair of Stress-Damaged Proteins

If the intracellular concentration of metal ions saturates the defense mechanisms discussed above then the plant will begin to suffer oxidative stress caused by the production of ROS and the inhibition of metal-dependent antioxidant enzymes (Schützendübel and Polle 2002). Under these circumstances, plants activate their antioxidant responses, including the induction of enzymes such as CAT and SOD and the production of non-enzymatic free radical scavengers. There are many examples of this process, such as the induction of APX and CAT in *Nicotiana plumbaginifolia* leaves exposed to excess Fe (Kampfenkel et al. 1995b), and the induction of *CAT3* in *B. juncea* plants exposed to Cd (Minglin et al. 2005). In pea plants, Cd causes the oxidation of CAT thus reducing its activity, so the plant responds by upregulating the transcription of the corresponding gene (Romero-Puertas et al. 2007). The tomato *chloronerva* mutant is NA-deficient and contains abnormally high levels of Fe and Cu in leaves, resulting in the activation of antioxidant enzymes such as cytosolic APX and Cu/Zn SOD (Pich and Scholz 1993; Herbick et al. 1996). SOD activity is also induced in tomato seedlings after prolonged Cd treatment (Dong et al. 2006). SOD activity also increases significantly in wheat leaves, but only following exposure to high levels of Cd, probably reflecting the accumulation of superoxide (Lin et al. 2007). Nevertheless, previous studies have shown that SOD activity is reduced in pea plants exposed to Cd toxicity (Romero-Puertas et al. 2007). An increase in *APX* mRNA is also observed in *Brassica napus* cotyledons subjected to toxic levels of Fe (Vansuyt et al. 1997). Several metals are able to induce Fe and Mn-SOD in plants (del Río et al. 1991).

The production of ROS is also countered by the activation of the ascorbic acid-GS scavenging system. In *Phaseolus vulgaris* and *Pisum sativum*, Cd treatment induces APX (Romero-Puertas et al. 2007). In addition, GR activity, another enzyme taking part in the cycle, is upregulated in roots and acts as a defence mechanism against Cd-generated oxidative stress (Yannarelli et al. 2007).

GS plays a key role in metal tolerance, because it can act as a ROS scavenger, metal chelator and as a substrate for PC biosynthesis (Krämer 2010). Indeed, the expression of the *E. coli* GSS gene *gshII* in *B. juncea* increased Cd tolerance in the seedlings and increased the capacity for Cd accumulation in adult plants (Zhu et al. 1999).

Heavy metals, in particular Cd, reduce the GSH/GSSG ratio and activate antioxidant enzymes such as SOD and GR (Romero-Puertas et al. 2007). GSH (Glu-Cys-Gly) is the major intracellular antioxidant inside the cell and is the precursor of PCs (Cobbett 2000). It can also form complexes with metal ions, particularly Cd (Wójcik and Tukiendorf 2011). GS synthesis is activated in response to heavy metal stress.

In common with other forms of abiotic stress, heavy metals induce the synthesis of stress-related proteins and signaling molecules, such as HSPs, SAPs, salicylic and abscisic acids, and ethylene. HSPs are found in all cells and are expressed not only in response to elevated temperatures but also to other stresses, including heavy metals, where they protect and repair proteins, and act as molecular chaperones to ensure correct folding and assembly (Vierling 1991). For example, heavy metals induce the expression of low molecular weight HSPs in rice (Tseng et al. 1993), and in cell culture of *S. vulgaris* and *Lycopersicon peruvianum* (Wollgiehn and Neumann 1999), the latter also producing the larger protein HSP70 (Neumann et al. 1994). HSP70 is localized in the nucleus and the cytoplasm, and is also found at the plasma membrane suggesting a protective role for membranes. These observations suggest that HSPs could have an important role in heavy metal response mechanism involving plasma membrane protection and in the repair of stress-damaged proteins.

The SAPs contain A20 or AN1 zinc finger domains (sometimes both) and, like HSPs, also respond to multiple abiotic stresses in plants, including cold, drought, salt, heavy metals, hypoxia, and wounding; they may function as transcriptional regulators or by direct protein–protein interactions (Dixit and Dhankher 2011). AtSAP10 is expressed in *A. thaliana* roots and is induced within 30 min by exposure to As, Cd, and Zn (Dixit and Dhankher 2011). Plants overexpressing AtSAP10 are metal tolerant, they accumulate Ni and Mn in the shoots and roots but there is no change in Zn content (Dixit and Dhankher 2011).

References

Allan DL, Jarrell WM (1989) Proton and copper adsorption to maize and soybean root cell walls. Plant Physiol 89:823–832

Andrés-Colás N, Sancenón V, Rodríguez-Navarro S, Mayo S, Thiele DJ, Ecker JR, Puig S, Peñarrubia L (2006) The *Arabidopsis* heavy metal P-type ATPase HMA5 interacts with metallochaperones and functions in copper detoxification of roots. Plant J 45:225–236

Arazi T, Sunkar R, Kaplan B, Fromm H (1999) A tobacco plasma membrane calmodulin-binding transporter confers Ni2+ tolerance and Pb2+ hypersensitivity in transgenic plants. Plant J 20:171–182

Ashrafi E, Alemzadeh A, Ebrahimi M, Ebrahimie E, Dadkhodaei N, Ebrahimi M (2011) Amino acid features of P1B-ATPase heavy metal transporters enabling small numbers of organisms to cope with heavy metal pollution. Bioinf Biol Insights 5:59–82

Assunção AGL, Da Costa Martins P, De Folter S, Vooijs R, Schat H, Aarts MGM (2001) Elevated expression of metal transporter genes in three accessions of the metal hyperaccumulator *Thlaspi caerulescens*. Plant Cell Environ 24:217–226

Axelsen KB, Palmgren MG (2001) Inventory of the Superfamily of P-Type Ion Pumps in *Arabidopsis*. Plant Physiol 126:696–706

Baxter I, Tchieu J, Sussman MR, Boutry M, Palmgren MG, Gribskov M, Harper JF, Axelsen KB (2003) Genomic comparison of P-Type ATPase ion pumps in *Arabidopsis* and rice. Plant Physiol 132:618–628

Bringezu K, Lichtenberger O, Leopold I, Neumann D (1999) Heavy metal tolerance of *Silene vulgaris*. J Plant Physiol 154:536–546

Callahan DL, Baker AJM, Kolev SD, Wedd AG (2006) Metal ion ligands in hyperaccumulating plants. J Biol Inorg Chem 11:2–12

Cataldo DA, McFadden KM, Garland TR, Wildung RE (1988) Organic constituents and complexation of nickel(II), iron(III), cadmium(II), and plutonium(IV) in soybean xylem exudates. Plant Physiol 86:734–739

Cherian GM, Chan HM (1993) Biological functions of metallothioneins: a review. In: Suzuki KT, Imura N, Kimura M (eds) Metallothionein III. Birkhauser, Basel

Cho U-H, Park J-O (2000) Mercury-induced oxidative stress in tomato seedlings. Plant Sci 156:1–9

Clemens S, Kim EJ, Neumann D, Schroeder JI (1999) Tolerance to toxic metals by a gene family of phytochelatin synthases from plants and yeast. EMBO J 15:3325–3333

Cobbett CS (2000) Phytochelatin biosynthesis and function in heavy-metal detoxification. Curr Opin Plant Biol 3:211–216

Cobbett CS, Goldsbrough P (2002) Phytochelatins and metallothioneins: roles in heavy metal detoxification and homeostasis. Annu Rev Plant Biol 53:159–182

Cohen CK, Fox TC, Garvin DF, Kochian LV (1998) The role of Iron-deficiency stress responses in stimulating heavy-metal transport in plants. Plant Physiol 116:1063–1072

Colangelo EP, Guerinot ML (2006) Put the metal to the petal: metal uptake and transport throughout plants. Curr Opin Plant Biol 9:322–330

Curie C, Alonso JM, Le Jean M, Ecker JR, Briat JF (2000) Involvement of NRAMP1 from *Arabidopsis thaliana* in iron transport. Biochem J 347:749–755

Curie C, Panaviene Z, Loulergue C, Dellaporta SL, Briat JF, Walker EL (2001) Maize *yellow stripe1* encodes a membrane protein directly involved in Fe(III) uptake. Nature 409:346–349

DalCorso G, Farinati S, Furini A (2010) Regulatory networks of cadmium stress in plants. Plant Signal Behav 5:663–667

Dat JF, Vandenabeele S, Vranova E, Van Montagu M, Inze D, Van Breusegem F (2000) Dual action of the active oxygen species during plant stress responses. Cell Mol Life Sci 57:779–795

De Knecht JA, van Dillen M, Koevoets PLM, Schat H, Verkleij JAC, Ernst WHO (1994) Phytochelatins in cadmium-sensitive and cadmium-tolerant Silene vulgaris. Chain length distribution and sulfide incorporation. Plant Physiol 104:255–261

Dedman DJ, Treffry A, Harrison PM (1992) Interaction of aluminium citrate with horse spleen ferritin. Biochem J 15:515–520

del Río LA, Sevilla F, Sandalio LM, Palma JM (1991) Nutritional effect and expression of SODs: induction and gene expression; diagnostics; prospective protection against oxygen toxicity. Free Radic Res Commun 2:819–827

DiDonato RJ Jr, Roberts LA, Sanderson T, Eisley RB, Walker EL (2004) Arabidopsis *Yellow Stripe-Like2* (*YSL2*): a metal-regulated gene encoding a plasma membrane transporter of nicotianamine–metal complexes. Plant J 39:403–414

Dietz KJ (1996) Functions and responses of the leaf apoplast under stress. Prog Bot 58:221–254

Dietz K-J, Baier M, Krämer U (1999) Free radicals and reactive oxygen species as mediators of heavy metal toxicity in plants. In: Prasad MNV, Hagemeyer J (eds) Heavy metal stress in plants: from molecules to ecosystems. Springer, Berlin

Dixit AR, Dhankher OP (2011) A novel stress-associated protein 'AtSAP10' from *Arabidopsis thaliana* confers tolerance to nickel, manganese, zinc, and high temperature stress. PLoS One 6:e20921

Dong J, Wu F, Zhang G (2006) Influence of cadmium on antioxidant capacity and four microelement concentrations in tomato seedlings (*Lycopersicon esculentum*). Chemosphere 64:1659–1666

Durrett TP, Gassmann W, Rogers EE (2007) The FRD3-mediated efflux of citrate into the root vasculature is necessary for efficient iron translocation. Plant Physiol 144:197–205

Ebbs S, Lau I, Ahner B, Kochian L (2002) Phytochelatin synthesis is not responsible for Cd tolerance in the Zn/Cd hyperaccumulator *Thlaspi caerulescens* (J. & C. Presl). Planta 214:635–640

Eide D, Broderius M, Fett J, Guerinot ML (1996) A novel iron-regulated metal transporter from plants identified by functional expression in yeast. Proc Natl Acad Sci U S A 93:5624–5628

Emery L, Whelan S, Hirschi KD, Pittman JK (2012) Phylogenetic analysis of Ca^{2+}/cation antiporter genes and insights into their evolution in plants. Front Plant Sci. doi:10.3389/fpls.2012.00001

Ernst WHO, Verkleij JAC, Schat H (1992) Metal tolerance in plants. Acta Bot Neerl 41:229–248

Evans KM, Gatehouse JA, Lindsay WP, Shi J, Tommey AM, Robinson NJ (1992) Expression of the pea metallothionein-like gene $PsMT_A$ in *Escherichia coli* and *Arabidopsis thaliana* and analysis of trace metal ion accumulation: implications for $PsMT_A$ function. Plant Mol Biol 20:1019–1028

Fürst P, Hu S, Hackett R, Hamer D (1988) Copper activates metallothionein gene transcription by altering the conformation of a specific DNA binding protein. Cell 55:705–717

García-Hernández M, Murphy A, Taiz L (1998) Metallothioneins 1 and 2 have distinct but overlapping expression patterns in Arabidopsis. Plant Physiol 118:387–397

Giritch A, Ganal M, Stephan UW, Baumlein H (1998) Structure, expression and chromosomal localization of the metallothionein-like gene family of tomato. Plant Mol Biol 37:701–714

Gong JM, Lee DA, Schroeder JI (2003) Long-distance root-to-shoot transport of phytochelatins and cadmium in *Arabidopsis*. Proc Natl Acad Sci U S A 100:10118–10123

Grill E, Loffler S, Winnacker E-L, Zenk MH (1989) Phytochelatins, the heavy-metal-binding peptides of plants are synthesized from glutathione by a specific γ-glutamylcysteine dipeptidyl transpeptidase (phytochelatin synthase). Proc Natl Acad Sci U S A 86:6838–6842

Grotz N, Fox T, Connolly E, Park W, Guerinot ML, Eide D (1998) Identification of a family of zinc transporter genes from *Arabidopsis* that respond to zinc deficiency. Proc Natl Acad Sci U S A 95:7220–7224

Guerinot ML (2000) The ZIP family of metal transporters. Biochim Biophys Acta 1465:190–198

Ha SB, Smith AP, Howden R, Dietrich WM, Bugg S, O'Connell MJ, Goldsbrough PB, Cobbett CS (1999) Phytochelatin synthase genes from Arabidopsis and the yeast *Schizosaccharomyces pombe*. Plant Cell 11:1153–1163

Haag-Kerwer A, Schäfer HJ, Heiss S, Walter C, Rausch T (1999) Cadmium exposure in *Brassica juncea* causes a decline in transpiration rate and leaf expansion without effect on photosynthesis. J Exp Bot 50:1827–1835

Hamer DH (1986) Metallothionein. Annu Rev Biochem 55:913–951

Harrison PM, Arosio P (1996) The ferritins: molecular properties, iron storage function and cellular regulation. Biochim Biophys Acta 275:161–203

Hartley-Whitaker J, Ainsworth G, Meharg AA (2001) Copper- and arsenate-induced oxidative stress in Holcus lanatus L. clones with differential sensitivity. Plant Cell and Environ 24:713–722

He C, Fong SH, Yang D, Wang GL (1999) BWMK1, a novel MAP kinase induced by fungal infection and mechanical wounding in rice. Mol Plant Microbe Interact 12:1064–1073

Herbik A, Giritch A, Horstmann C, Becker R, Balzer HJ, Bäumlein H, Stephan UW (1996) Iron and copper nutrition-dependent changes in protein expression in a tomato wild type and the nicotianamine-free mutant *chloronerva*. Plant Physiol 111:533–540

Hocking PJ, Pate JS (1977) Mobilization of minerals to developing seeds of legumes. Ann Bot 41:1259–1278

Howden R, Goldsbrough PB, Andersen CR, Cobbett CS (1995) Cadmium-sensitive, cad1 mutants of Arabidopsis thaliana are phytochelatin deficient. Plant Physiol 107:1059–1066

Hubel F, Beck E (1996) Maize root phytase (purification, characterization, and localization of enzyme activity and its putative substrate). Plant Physiol 112:1429–1436

Hussain D, Haydon MJ, Wang Y, Wong E, Sherson SM, Young J, Camakaris J, Harper JF, Cobbett CS (2004) P-Type ATPase heavy metal transporters with roles in essential zinc homeostasis in *Arabidopsis*. Plant Cell 16:1327–1339

Jonak C, Kiegerl S, Ligterink W, Barker PJ, Huskisson NS, Hirt H (1996) Stress signaling in plants: a mitogen-activated protein kinase pathway is activated by cold and drought. Proc Natl Acad Sci U S A 93:11274–11279

Jonak C, Okrész L, Bögre L, Hirt H (2002) Complexity, cross talk and integration of plant MAP kinase signalling. Curr Opin Plant Biol 5:415–424

Jonak C, Nakagami H, Hirt H (2004) Heavy metal stress. Activation of distinct mitogen-activated protein kinase pathways by copper and cadmium. Plant Physiol 136:3276–3283

Kampfenkel K, Kushnir S, Babiychuk E, Inzé D, Van Montagu M (1995a) Molecular characterization of a putative *Arabidopsis thaliana* copper transporter and its yeast homologue. J Biol Chem 270:28479–28486

Kampfenkel K, Van Montagu M, Inze D (1995b) Effects of iron excess on *Nicotiana plumbaginifolia* plants (Implications to oxidative stress). Plant Physiol 107:725–735

Kawachi M, Kobae Y, Mimura T, Maeshima M (2008) Deletion of a histidine-rich loop of AtMTP1, a vacuolar Zn^{2+}/H^+ antiporter of *Arabidopsis thaliana*, stimulates the transport activity. J Biol Chem 283:8374–8383

Kim DY, Bovet L, Maeshima M, Martinoia E, Lee Y (2007) The ABC transporter AtPDR8 is a cadmium extrusion pump conferring heavy metal resistance. Plant J 50:207–218

Knight H (1999) Calcium signaling during abiotic stress in plants. Int Rev Cytol 195:269–324

Kobae Y, Sekino T, Yoshioka H, Nakagawa T, Martinoia E, Maeshima M (2006) Loss of AtPDR8, a plasma membrane ABC transporter of *Arabidopsis thaliana*, causes hypersensitive cell death upon pathogen infection. Plant Cell Physiol 47:309–318

Korenkov V, Park SH, Cheng NH, Sreevidya C, Lachmansingh J, Morris J, Hirschi K, Wagner GJ (2007) Enhanced Cd^{2+} selective root-tonoplast-transport in tobaccos expressing *Arabidopsis* cation exchangers. Planta 225:403–411

Korshunova YO, Eide D, Clark WG, Guerinot ML, Pakrasi HB (1999) The IRT1 protein from *Arabidopsis thaliana* is a metal transporter with a broad substrate range. Plant Mol Biol 40:37–44

Krämer U (2010) Metal hyperaccumulation in plants. Annu Rev Plant Biol 61:28.1–28.18

Krämer U, Cotter-Howells JD, Charnock JM, Baker AJM, Smith JAC (1996) Free histidine as a metal chelator in plants that accumulate nickel. Nature 379:635–638

Krämer U, Talke IN, Hanikenne M (2007) Transition metal transport. FEBS Lett 581:2263–2272

Lanquar V, Lelièvre F, Bolte S, Hamès C, Alcon C, Neumann D, Vansuyt G, Curie C, Schröder A, Krämer U, Barbier-Brygoo H, Thomine S (2005) Mobilization of vacuolar iron by AtNRAMP3 and AtNRAMP4 is essential for seed germination on low iron. EMBO J 24:4041–4051

Le Jean M, Schikora A, Mari M, Briat JF, Curie C (2005) A loss-of-function mutation in *AtYSL1* reveals its role in iron and nicotianamine seed loading. Plant J 44:769–782

Lee S, Moon JS, Ko TS, Petros D, Goldsbrough PB, Korban SS (2003) Overexpression of *Arabidopsis* phytochelatin synthase paradoxically leads to hypersensitivity to cadmium stress. Plant Physiol 131:656–663

Lee J, Shim D, Song WY, Hwang I, Lee Y (2004) *Arabidopsis* metallothioneins 2a and 3 enhance resistance to cadmium when expressed in *Vicia faba* guard cells. Plant Mol Biol 54:805–815

Leea J, Reevesa RD, Brooksa RR, Jaffréb T (1977) Isolation and identification of a citrato-complex of nickel from nickel-accumulating plants. Phytochemistry 16:1503–1505

Li Z-S, Lu Y-P, Zhen R-G, Szczypka M, Thiele DJ, Rea PA (1997) A new pathway for vacuolar cadmium sequestration in *Saccharomyces cerevisiae*: YCF1-catalyzed transport of bis(gluta-thionato)cadmium. Proc Natl Acad Sci U S A 94:42–47

Lin R, Wang X, Luo Y, Du W, Guo H, Yin D (2007) Effects of soil cadmium on growth, oxidative stress and antioxidant system in wheat seedlings (*Triticum aestivum* L.). Chemosphere 69:89–98

Lu YP, Li ZS, Rea PA (1997) *AtMRP1* gene of Arabidopsis encodes a glutathione *S*-conjugate pump: isolation and functional definition of a plant ATP-binding cassette transporter gene. Proc Natl Acad Sci U S A 94:8243–8248

Lu YP, Li ZS, Drozdowicz YM, Hortensteiner S, Martinoia E, Rea PA (1998) AtMRP2, an *Arabidopsis* ATP binding cassette transporter able to transport glutathione *S*-conjugates and chlorophyll catabolites: functional comparisons with AtMRP1. Plant Cell 10:267–282

Maas FM, van de Wetering DAM, van Beusichem ML, Bienfait HF (1988) Characterization of phloem iron and its possible role in the regulation of Fe-efficiency reactions. Plant Physiol 87:167–171

MacDiarmid CW, Gaither LA, Eide D (2000) Zinc transporters that regulate vacuolar zinc storage in *Saccharomyces cerevisiae*. EMBO J 19:2845–2855

Maksymiec (2007) Signaling responses in plants to heavy metal stress. Acta Physiol Plant 29:177–187

Maksymiec W, Krupa Z (2006) The effects of short-term exposition to Cd, excess Cu ions and jasmonate on oxidative stress appearing in *Arabidopsis thaliana*. Environ Exp Bot 57:187–194

Maksymiec W, Wianowska D, Dawidowicz AL, Radkiewicz S, Mardarowicz M, Krupa Z (2005) The level of jasmonic acid in *Arabidopsis thaliana* and *Phaseolus coccineus* plants under heavy metal stress. J Plant Physiol 162:1338–1346

Marschner H (1995) Mineral nutrition of higher plants 2nd edn. Academic Press, London

Martinoia E, Grill E, Tommasini R, Kreuz K, Amrhein N (1993) ATP-dependent glutathione *S*-conjugate 'export' pump in the vacuolar membrane of plants. Nature 364:247–249

Mäser P, Thomine S, Schroeder JI, Ward JM, Hirschi K, Sze H, Talke IN, Amtmann A, Maathuis FJM, Sanders D, Harper JF, Tchieu J, Gribskov M, Persans MW, Salt DE, Kim SA, Guerinot ML (2001) Phylogenetic Relationships within Cation Transporter Families of *Arabidopsis*. Plant Physiol 126:1646–1667

Mathys W (1977) The role of malate, oxalate, and mustard oil glucosides in the evolution of zinc-resistance in herbage plants. Physiol Plantarum 40:130–136

Meharg AA, Macnair MR (1992) Genetic correlation between arsenate tolerance and the rate of influx of arsenate and phosphate in *Holcus lanatus* L. Heredity 69:336–341

Mei H, Cheng NH, Zhao J, Park S, Escareno RA, Pittman JK, Hirschi KD (2009) Root development under metal stress in *Arabidopsis thaliana* requires the H^+/cation antiporter CAX4. New Phytol 183:95–105

Metwally A, Finkemeier I, Georgi M, Dietz K-J (2003) Salicylic acid alleviates the cadmium toxicity in barley seedlings. Plant Physiol 132:272–281

MikuS M, Bobak M, Lux A (1992) Structure of protein bodies and elemental composition of phytin from dry germ of maize (*Zea mays* L.). Bot Acta 105:26–33

Mills RF, Krijger GC, Baccarini PJ, Hall JL, Williams LE (2003) Functional expression of AtHMA4, a P_{1B}-type ATPase of the Zn/Co/Cd/Pb subclass. Plant J 35:164–176

Mills RF, Francini A, Ferreira da Rocha PSC, Baccarini PJ, Aylett M, Krijger GC, Williams LE (2005) The plant P1B-type ATPase AtHMA4 transports Zn and Cd and plays a role in detoxification of transition metals supplied at elevated levels. FEBS Lett 579:783–791

Minglin L, Yuxiu Z, Tuanyao C (2005) Identification of genes up-regulated in response to Cd exposure in *Brassica juncea* L. Gene 363:151–158

Montanini B, Blaudez D, Jeandroz S, Sanders D, Chalot M (2007) Phylogenetic and functional analysis of the cation diffusion facilitator (CDF) family: improved signature and prediction of substrate specificity. BMC Genomics 8:107. doi:10.1186/1471-2164-8-107

Morel M, Crouzet J, Gravot A, Auroy P, Leonhardt N, Vavasseur A, Richaud P (2009) AtHMA3, a P1B-ATPase allowing Cd/Zn/Co/Pb vacuolar storage in Arabidopsis. Plant Physiol 149:894–904

Murgia I, Briat JF, Tarantino D, Soave C (2001) Plant ferritin accumulates in response to photoinhibition but its ectopic overexpression does not protect against photoinhibition. Plant Physiol Bioc 39:797–805

Murgia I, Delledonne M, Soave C (2002) Nitric oxide mediates iron-induced ferritin accumulation in Arabidopsis. Plant J 30:521–528

Neumann D, Lichtenberger O, Günther D, Tschiersch K, Nover L (1994) Heat-shock proteins induce heavy-metal tolerance in higher plants. Planta 194:360–367

Nevo Y, Nelson N (2006) The NRAMP family of metal-ion transporters. Biochim Biophys Acta 1763:609–620

Nies DH (1992) Resistance to cadmium, cobalt, zinc, and nickel in microbes. Plasmid 27:17–28

Nishida S, Mizuno T, Obata H (2008) Involvement of histidine-rich domain of ZIP family transporter TjZNT1 in metal ion specificity. Plant Physiol Biochem 46:601–606

Nishida S, Tsuzuki C, Kato A, Aisu A, Yoshida J, Mizuno T (2011) AtIRT1, the primary iron-uptake rransporter in the root, mediates excess nickel accumulation in *Arabidopsis thaliana*. Plant Cell Physiol 52:1433–1442

Ortiz DF, Kreppel L, Speiser DM, Scheel G, McDonald G, Ow DW (1992) Heavy metal tolerance in the fission yeast requires an ATP-binding cassette-type vacuolar membrane transporter. EMBO J 11:3491–3499

Ortiz DF, Ruscitti T, McCue KF, Ow DW (1995) Transport of metal-binding peptides by HMT1, a fission Yeast ABC-type vacuolar membrane protein. J Biolg Chem 270:4721–4728

Palmiter RD, Findley SD (1995) Cloning and functional characterization of a mammalian zinc transporter that confers resistance to zinc. EMBO J 14:639–649

Peer WA, Mamoudian M, Lahner B, Reeves RD, Murphy AS, Salt DE (2003) Identifying model metal hyperaccumulating plants: germplasm analysis of 20 *Brassicaceae* accessions from a wide geographical area. New Phytol 159:421–430

Peiter E, Montanini B, Gobert A, Pedas P, Husted S, Maathuis FJM, Blaudez D, Chalot M, Sanders D (2007) A secretory pathway-localized cation diffusion facilitator confers plant manganese tolerance. Proc Natl Acad Sci U S A 104:8532–8537

Peleg Z, Blumwald E (2011) Hormone balance and abiotic stress tolerance in crop plants. Curr Opin Plant Biol 14:290–295

Pell EJ, Schlagnhaufer CD, Arteca RN (1997) Ozone-induced oxidative stress: Mechanisms of action and reaction. Physiol Plantarum 100:264–273

Pich A, Scholz G (1993) The relationship between the activity of various iron-containing and iron-free enzymes and the presence of nicotianamine in tomato seedlings. Physiol Plant 88:172–178

Pich A, Scholz G (1996) Translocation of copper and other micronutrients in tomato plants (*Lycopersicon esculentum* Mill.): nicotianamine-stimulated copper transport in the xylem. J Exp Bot 47:41–47

Pich A, Hillmer S, Manteuffel R, Scholz G (1997) First immunohistochemical localization of the endogenous Fe^{2+-}chelator nicotianamine. J Exp Bot 48:759–767

Pomponi M, Censi V, Di Girolamo V et al (2006) Overexpression of *Arabidopsis* phytochelatin synthase in tobacco plants enhances Cd^{2+} tolerance and accumulation but not translocation to the shoot. Planta 223:180–190

Price DJ, Joshi JG (1982) Ferritin. Binding of beryllium and other divalent metal ions. J Biol Chem 258:10873–10880

Puig S, Thiele DJ (2002) Molecular mechanisms of copper uptake and distribution. Curr Opin Chem Biol 6:171–180

Rakwal R, Tamogami S, Kodama O (1996) Role of jasmonic acid as a signaling molecule in copper chloride-elicited rice phytoalexin production. Biosci Biotech Bioch 60:1046–1048

Rauser WE (1995) Phytochelatins and related peptides. Structure, biosynthesis, and function. Plant Physiol 109:1141–1149

Rauser WE (1999) Structure and function of metal chelators produced by plants. The case for organic acids, amino acids, phytin and metallothioneins. Cell Biochem Biophys 31:19–48

Ravet K, Touraine B, Boucherez J, Briat J-F, Gaymard F, Cellier F (2009) Ferritins control interaction between iron homeostasis and oxidative stress in *Arabidopsis*. Plant J 57:400–412

Roberts LA, Pierson AJ, Panaviene Z, Walker EL (2004) Yellow Stripe1. Expanded roles for the maize iron-phytosiderophore transporter. Plant Physiol 135:112–120

Robinson NJ, Tommey AM, Kuske C, Jackson PJ (1993) Plant metallothioneins. Biochem J 295:1–10

Robinson NJ, Wilson JR, Turner JS (1996) Expression of the type 2 metallothionein-like gene MT2 from *Arabidopsis thaliana* in Zn^{2+}-metallothionein-deficient *Synechococcus* PCC 7942: putative role for MT2 in Zn^{2+} metabolism. Plant Mol Biol 30:1169–1179

Romero-Puertas MC, Corpas FJ, Rodriguez-Serrano M, Gomez M, del Río LA, Sandalio LM (2007) Differential expression and regulation of antioxidative enzymes by cadmium in pea plants. J Plant Physiol 164:1346–1357

Salt DE, Rauser WE (1995) MgATP-dependent transport of phytochelatins across the tonoplast of oat roots. Plant Physiol 107:1293–1301

Salt DE, Wagner GJ (1993) Cadmium transport across tonoplast of vesicles from oat roots. Evidence for a Cd^{2+}/H^+ antiport activity. J Biol Chem 268:12297–12302

Salt DE, Thurman DA, Tomsett AB, Sewell AK (1989) Copper phytochelatins of Mimulus guttatus. P Royl Soc B-Biol Sci 236:79–89

Salt DE, Smith RD, Raskin I (1998) Phytoremediation. Annu Rev Plant Phys 49:643–668

Salt DE, Kato N, Krämer U, Smith RD, Raskin I (2000) The role of root exudates in nickel hyperaccumulation and tolerance in accumulator and nonaccumulator species of *Thlaspi*. In: Terry N, Banuelos G (eds) Phytoremediation of contaminated soil and water. CRC Press LLC, Boca Raton

Sancenón V, Puig S, Mateu-Andrés I, Dorcey E, Thiele DJ, Peñarrubia L (2004) The *Arabidopsis* copper transporter COPT1 functions in root elongation and pollen development. J Biol Chem 279:15348–15355

Savino G, Briat JF, Lobréaux S (1997) Inhibition of the iron-induced *ZmFer1* maize ferritin gene expression by antioxidants and serine/threonine phosphatase inhibitors. J Biol Chem 272:33319–33326

Schaaf G, Ludewig U, Erenoglu BE, Mori S, Kitahara T, von Wirén N (2004) ZmYS1 functions as a proton-coupled symporter for phytosiderophore- and nicotianamine-chelated metals. J Biol Chem 279:9091–9096

Schaaf G, Schikora A, Häberle J, Vert G, Ludewig U, Briat JF, Curie C, von Wirén N (2005) A putative function for the *Arabidopsis* Fe-Phytosiderophore transporter homolog AtYSL2 in Fe and Zn homeostasis. Plant Cell Physiol 46:762–774

Schat H, Kalff MMA (1992) Are phytochelatins involved in differential metal tolerance or do they merely reflect metal-imposed strain? Plant Physiol 99:1475–1480

Scholz G, Schlesier G, Seifert K (1985) Effect of nicotianamine on iron uptake by the tomato mutant 'chloronerva'. Physiol Plant 63:99–104

Schützendübel A, Polle A (2002) Plant responses to abiotic stresses: heavy metal-induced oxidative stress and protection by mycorrhization. J Exp Bot 53:1351–1365

Sekhar K, Priyanka B, Reddy VD, Rao KV (2011) Metallothionein 1 (CcMT1) of pigeonpea (*Cajanus cajan*, L.) confers enhanced tolerance to copper and cadmium in *Escherichia coli* and *Arabidopsis thaliana*. Environ Exp Bot 72:131–139

Shaul O, Hilgemann DW, de-Almeida-Engler J J, Van Montagu M, Inzé D, Galili G (1999) Cloning and characterization of a novel Mg^{2+}/H^+ exchanger. EMBO J 18:3973–3980

Shojima S, Nishizawa NK, Fushiya S, Nozoe S, Irifune T, Mori S (1990) Biosynthesis of phytosiderophores: in vitro biosynthesis of 2′-deoxymugineic acid from L-methionine and nicotianamine. Plant Physiol 93:1497–1503

Silver S (1996) Bacterial resistance to toxic metal ions—a review. Gene 179:9–19

Skórzyńska-Polit E, Tukendorf A, Selstam E, Baszyński T (1998) Calcium modifies Cd effect on runner bean plants. Environ Exp Bot 40:275–286

Steffens JC (1990) The heavy metal-binding peptides of plants. Annu Rev Plant Phys 41:553–575

Stephan UW, Scholz G (1993) Nicotianamine: mediator of transport of iron and heavy metals in the phloem? Physiol Plantarum 88:522–529

Stephan UW, Schmidke I, Pich A (1994) Phloem translocation of Fe, Cu, Mn, and Zn in *Ricinus* seedlings in relation to the concentrations of nicotianamine, an endogenous chelator of divalent metal ions, in different seedling parts. Plant Soil 165:181–188

Stephan UW, Schmidke I, Stephan VW, Scholz G (1996) The nicotianamine molecule is made-to-measure for complexation of metal micronutrients in plants. Biometals 9:84–90

Taylor KC, Albrigo LG, Chase CD (1988) Zinc complexation in the phloem of blight-affected citrus. J Am Soc Hortic Sci 113:407–411

Thomine S, Wang R, Ward JM, Crawford NM, Schroeder JI (2000) Cadmium and iron transport by members of a plant metal transporters family in *Arabidopsis* with homology to Nramp genes. Proc Natl Acad Sci U S A 97:4991–4996

Thomine S, Lelievre F, Debarbieux E, Schroeder JI, Barbier-Brygoo H (2003) AtNRAMP3, a multispecific vacuolar metal transporter involved in plant responses to iron deficiency. Plant J 34:685–695

Thumann J, Grill E, Winnacker EL, Zenk MH (1991) Reactivation of metal-requiring apoenzymes by phytochelatin-metal complexes. FEBS Lett 284:66–69

Tiffin LO (1970) Translocation of iron citrate and phosphorus in xylem exudate of soybean. Physiol 45:280–283

Tommasini R, Vogt E, Fromenteau M, Hörtensteiner S, Matile P, Amrhein N, Martinoia E (1998) An ABC-transporter of *Arabidopsis thaliana* has both glutathione-conjugate and chlorophyll catabolite transport activity. Plant J 13:773–780

Tommey AM, Shi J, Lindsay WP, Urwin PE, Robinson NJ (1991) Expression of the pea gene *PsMT_A* in *E. coli*. Metal binding properties of the expressed protein. FEBS Lett 292:48–52

Treeby M, Marschner H, Römheld V (1989) Mobilization of iron and other micronutrient cations from a calcareous soil by plant-borne, microbial, and synthetic metal chelators. Plant Soil 114:217–226

Tseng TS, Tzeng SS, Yeh CH, Chang FC, Chen YM, Lin CY (1993) The heat-shock response in rice seedlings: isolation and expression of cDNAs that encode class-I low-molecular-weight heat-shock proteins. Plant Cell Physiol 34:165–168

van der Zaal BJ, Neuteboom LW, Pinas JE, Chardonnens AN, Schat H, Verkleij JAC, Hooykaas PJJ (1999) Overexpression of a novel *Arabidopsis* gene related to putative zinc-transporter genes from animals can lead to enhanced zinc resistance and accumulation. Plant Physiol 119:1047–1055

van Steveninck RFM, Babare A, Fernando DR, Steveninck ME (1993) The binding of zinc in root cells of crop plants by phytic acid. Plant Soil 155–156:525–528

Vansuyt G, Lopez F, Inzé D, Briat J-F, Fourcroy P (1997) Iron triggers a rapid induction of ascorbate peroxidase gene expression in *Brassica napus*. FEBS Lett 410:195–200

Vatamaniuk OK, Mari S, Lu Y-P, Rea PA (1999) AtPCS1, a phytochelatin synthase from *Arabidopsis*: Isolation and in vitro reconstitution. Proc Natl Acad Sci U S A 96:7110–7115

Verret F, Gravot A, Auroy P, Leonhardt N, David P, Nussaume L, Vavasseur A, Richaud P (2004) Overexpression of AtHMA4 enhances root-to-shoot translocation of zinc and cadmium and plant metal tolerance. FEBS Lett 576:306–312

Verret F, Gravot A, Auroy P, Preveral S, Forestier C, Vavasseur A, Richaud P (2005) Heavy metal transport by AtHMA4 involves the N-terminal degenerated metal binding domain and the C-terminal His_{11} stretch. FEBS Lett 579:1515–1522

Vert G, Grotz N, Dédaldéchamp F, Gaymard F, Guerinot ML, Briat J-F, Curie C (2002) IRT1, an *Arabidopsis* transporter essential for iron uptake from the soil and for plant growth. Plant Cell 14:1223–1233

Vierling E (1991) The roles of heat shock proteins in plants. Ann Rev Plant Phys 42:579–620

Vögeli-Lange R, Wagner GJ (1990) Subcellular localization of cadmium and cadmium-binding peptides in tobacco leaves. Implication of a transport function for cadmium-binding peptides. Plant Physiol 92:1086–1093

Wang J, Evangelou BP, Nielsen MT (1992) Surface chemical properties of purified root cell walls from two tobacco genotypes exhibiting different tolerance to manganese toxicity. Plant Physiol 100:496–501

Wenzel WW, Bunkowski M, Puschenreiter M, Horak O (2003) Rhizosphere characteristics of indigenously growing nickel hyperaccumulator and excluder plants on serpentine soil. Environ Pollut 123:131–138

Wiersma D, Van Goor BJ (1979) Chemical forms of nickel and cobalt in phloem of *Ricinus communis*. Physiol Plantarum 45:440–442

Williams LE, Pittman JK, Hall JL (2000) Emerging mechanisms for heavy metal transport in plants. Biochim Biophys Acta 77803:1–23

Wise RR, Naylor AW (1988) Stress ethylene does not originate directly from lipid peroxidation during chilling- enhanced photooxidation. J Plant Physiol 133:62–66

Wójcik M, Tukiendorf A (2011) Glutathione in adaptation of *Arabidopsis thaliana* to cadmium stress. Biol plantarum 55:125–132

Wollgiehn R, Neumann D (1999) Metal stress response and tolerance of cultured cells from *Silene vulgaris* and *Lycopersicon peruvianum*: role of heat stress proteins. J Plant Physiol 154:547–553

Wong CKE, Cobbett CS (2009) HMA P-type ATPases are the major mechanism for root-to-shoot Cd translocation in *Arabidopsis thaliana*. New Phytol 181:71–78

Xiang C, Oliver DJ (1998) Glutathione metabolic genes coordinately respond to heavy metals and jasmonic acid in Arabidopsis. Plant Cell 10:1539–1550

Xiang C, Werner BL, Christensen EM, Oliver DJ (2001) The biological functions of glutathione revisited in *Arabidopsis* transgenic plants with altered glutathione levels. Plant Physiol 126:564–574

Yang T, Poovaiah BW (2003) Calcium/calmodulin-mediated signal network in plants. Trends Plant Sci 8:505–512

Yannarelli GG, Fernàndez-Alvarez AJ, Santa-Cruz DM, Tomaro ML (2007) Glutathione reductase activity and isoforms in leaves and roots of wheat plants subjected to cadmium stress. Phytochemistry 68:505–512

Yeh C-M, Hsiao L-J, Huang H-J (2004) Cadmium activates a mitogen-activated protein kinase gene and MBP kinases in rice cell. Plant Cell Physiol 45:1306–1312

Zenk MH (1996) Heavy metal detoxification in higher plants—a review. Gene 179:21–30

Zhao H, Eide D (1996a) The yeast *ZRT1* gene encodes the zinc transporter protein of a high-affinity uptake system induced by zinc limitation. Proc Natl Acad Sci U S A 93:2454–2458

Zhao H, Eide D (1996b) The *ZRT2* gene encodes the low affinity zinc transporter in *Saccharomyces cerevisiae*. J Biol Chem 271:23203–23210

Zhigang A, Cuijie L, Yuangang Z, Yejie D, Wachter A, Gromes R, Rausch T (2006) Expression of BjMT2, a metallothionein 2 from *Brassica juncea*, increases copper and cadmium tolerance in *Escherichia coli* and *Arabidopsis thaliana*, but inhibits root elongation in *Arabidopsis thaliana* seedlings. J Exp Bot 57:3575–3582

Zhou J, Goldsbrough PB (1994) Functional homologs of fungal metallothionein genes from *Arabidopsis*. Plant Cell 6:875–884

Zhu YL, Pilon-Smits EAH, Tarun AS, Weber SU, Jouanin L, Terry N (1999) Cadmium tolerance and accumulation in Indian mustard is enhanced by overexpressing gamma-glutamylcysteine synthetase. Plant Physiol 121:1169–1177

Zimeri AM, Dhankher OP, McCaig B, Meagher RB (2005) The plant MT1 metallothioneins are stabilized by binding cadmium and are required for cadmium tolerance and accumulation. Plant Mol Biol 58:839–855

Chapter 3
Plants that Hyperaccumulate Heavy Metals

Elisa Fasani

Abstract Heavy metal hyperaccumulators are plants that can tolerate and accumulate extremely high concentrations of metals in their shoots. This reflects the enhancement of physiological processes such as metal uptake, mobilization, translocation, and detoxification by chelation and vacuolar sequestration. Hyperaccumulation occurs in approximately 500 taxa of angiosperms and is particularly common among the Brassicaceae. Several candidate genes have been proposed as determinants of heavy metal hyperaccumulation. They predominantly encode transporters involved in metal translocation and storage, and also chelators and genes involved in stress responses.

Keywords Hyperaccumulator · Elemental defense · Metal transporter · Metal ligand

3.1 Defining Hyperaccumulator Plants

The first plant species reported to accumulate extremely high levels of metals was *Alyssum bertolonii*, whose Ni content was greater than 1 mg g^{-1} dry weight (Minguzzi and Vergnano 1948). However, the term "hyperaccumulator" was coined only in 1976 to describe plants whose shoot metal concentration is some orders of magnitude higher than adjacent plants (Jaffrè et al. 1976). This definition implies high rates of metal uptake in roots, translocation, and accumulation in

E. Fasani (✉)
Department of Biotechnology, University of Verona,
Strada Le Grazie 15, 37134 Verona, Italy
e-mail: elisa.fasani@univr.it

A. Furini (ed.), *Plants and Heavy Metals*, SpringerBriefs in Biometals,
DOI: 10.1007/978-94-007-4441-7_3, © Fasani 2012

Table 3.1 Hyperaccumulation thresholds for the most relevant heavy metals, in comparison with the average content in plant tissues and toxicity levels

Element	Average range in plant tissues (mg/kg dw)[a]	Critical toxicity level (mg/kg dw)[b]	Threshold for hyper-accumulators (mg/kg dw)[a,b]
As	<0.01–4[c]	<2–80	>1,000
Cd	0.03–0.5	6–10	>100
Co	0.01–3[d]	0.4-several	>1,000
Cu	2–20	20–30	>1,000
Cr	0.2[e]	0.2–1	>1,000
Pb	0.1–5	0.6–28	>1,000
Mn	1–700	200–3,500	>10,000
Hg	0.005–0.2	0.001–5[f]	>1,000
Ni	0.4–4	10–50	>1,000
Se[h]	0.01–0.2	3–100	>1,000
Tl	0.1–1.5[g]	20	>1,000
Zn	15–150	100–300	>10,000

[a] *from* Maestri et al. (2010)
[b] *from* Krämer (2010)
[c] *from* National Research Council, Committee on Medical and Biological Effects of Environmental Pollutants (1977)
[d] *from* Sillanpää and Jansson (1992)
[e] *from* National Research Council, Committee on Biologic Effects of Atmospheric Pollutants (1974)
[f] *from* Patra et al. (2004)
[g] *from* Kazantzis (2000)
[h] Although Se is not a heavy metal, it is included because a number of Se hyperaccumulators have also been discovered

shoots, thus excluding plants which accumulate metals in the roots alone (Maestri et al. 2010). Inevitably, hyperaccumulation implies hypertolerance, i.e., the ability of the plants to detoxify heavy metals stored in aerial tissues (Krämer 2010). Thresholds have been set for different metals and metalloids to define plants as hyperaccumulators (Table 3.1).

Hyperaccumulation is an extreme trait that has evolved many times but is relatively uncommon in terrestrial higher plants. The metal hyperaccumulators identified thus far belong to approximately 500 taxa, accounting for 0.2% of all angiosperms (Baker et al. 2000; Krämer 2010). The number of hyperaccumulator taxa discovered for the main heavy metals is shown in Table 3.2. Most of the known hyperaccumulators are biennial or short-lived perennial herbs, shrubs or small trees. They are mainly endemic to metal-rich soils and are often unable to compete with other species in non-selective soils, possibly due to the higher metabolic costs of metal accumulation and detoxification (Baker et al. 2000). The hyperaccumulation trait is particularly well represented among the Brassicaceae. A phylogenetic tree of the Brassicaceae showing the positions of the main hyperaccumulator species is shown in Fig. 3.1.

Ni hyperaccumulation is the most common trait, reflecting the large number of Ni-enriched serpentine soils worldwide, in particular in the Mediterranean area

Table 3.2 Number of hyperaccumulator plants discovered to date for relevant heavy metals, as reported by Krämer (2010), with modification regarding Cd hyperaccumulator species cited by Liu et al. (2009)

Element	Taxa (no.)	Families (no.)
As	15	2
Cd	8	5
Co	(26)[a]	(11)
Cu	(35)	(15)
Pb	(14)	(7)
Mn	10	6
Ni	390	42
Se[b]	20	7
Tl	1	1
Zn	15	6

[a] Parentheses indicate that overreporting may have occurred due to contamination with soil particles or minerals
[b] Although Se is not a heavy metal, it is included because a number of Se hyperaccumulators have also been discovered

and in New Caledonia. Among the nearly 400 known metal hyperaccumulator species, approximately 25% are from the families Brassicaceae and Euphorbiaceae (Krämer 2010). Evident from Fig. 3.1, Ni hyperaccumulation has evolved independently six times in the Brassicaceae (Krämer 2010) and occurs most frequently in the genus *Alyssum* (Baker et al. 2000), almost exclusively in the section Odontarrhena (Krämer 2010). As far as Zn is concerned, the majority of hyperaccumulators belong to the Brassicaceae, with probably three independent evolutionary events. Zn hyperaccumulation tends to correlate with Cd and Pb accumulation because these metals share similar chemical properties (Krämer 2010). The only known Cd hyperaccumulator species outside the Brassicaceae are *Viola baoshanensis* (Violaceae; Liu et al. 2004), *Salsola kali* (Chenopodiaceae; de la Rosa et al. 2004), *Sedum alfredii* (Crassulaceae; Deng et al. 2007) and *Phytolacca americana* (Phytolaccaceae; Liu et al. 2009). Finally, As hyperaccumulation has been reported in only two angiosperm species, both belonging to the Brassicaceae (Karimi et al. 2009). Interestingly, the only other known As hyperaccumulators are some fern species from the genus *Pteris* (Zhao et al. 2002), among which the most studied is *P. vittata* (Wang et al. 2002). Species that hyperaccumulate other metals, such as Se (Reeves and Baker 2000) and Pb (Baker et al. 2000), have also been identified.

Two model species for hyperaccumulation, *Arabidopsis* (formerly *Cardaminopsis*) *halleri* and *Noccaea* (formerly *Thlaspi*) *caerulescens*, are particularly suitable for genetic analysis thanks to their strong similarity and extensive synteny with *A. thaliana*.

Arabidopsis halleri is a self-incompatible perennial diploid species that can tolerate and hyperaccumulate Zn and Cd. It shares 94% sequence identity with *A. thaliana* within coding regions (Clauss and Koch 2006) and appears to have

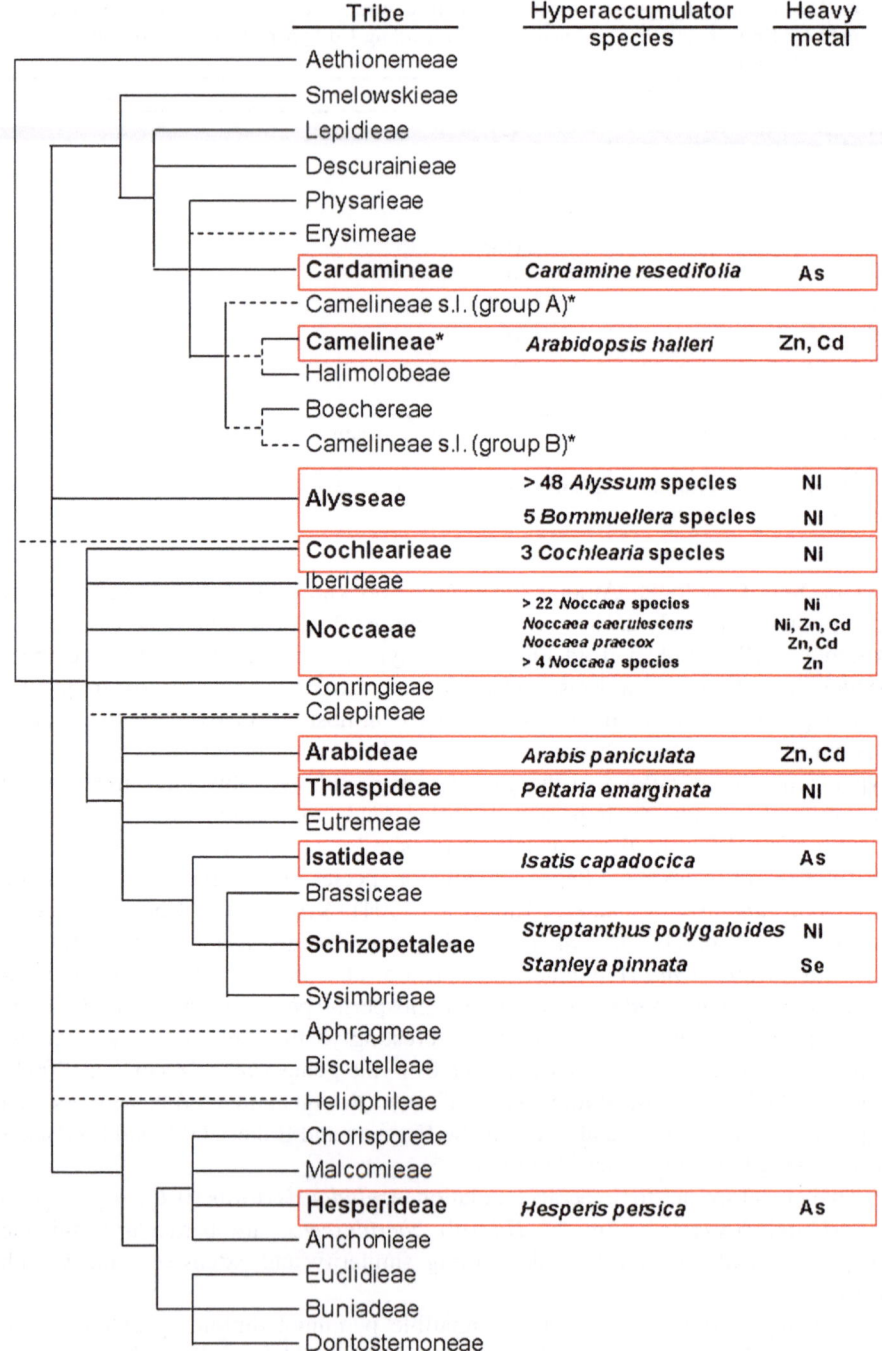

Tribe	Hyperaccumulator species	Heavy metal
Aethionemeae		
Smelowskieae		
Lepidieae		
Descurainieae		
Physarieae		
Erysimeae		
Cardamineae	*Cardamine resedifolia*	**As**
Camelineae s.l. (group A)*		
Camelineae*	*Arabidopsis halleri*	**Zn, Cd**
Halimolobeae		
Boechereae		
Camelineae s.l. (group B)*		
Alysseae	> 48 *Alyssum* species	**Ni**
	5 *Bornmuellera* species	**Ni**
Cochlearieae	3 *Cochlearia* species	**Ni**
Iberideae		
Noccaeae	> 22 *Noccaea* species	**Ni**
	Noccaea caerulescens	**Ni, Zn, Cd**
	Noccaea praecox	**Zn, Cd**
	> 4 *Noccaea* species	**Zn**
Conringieae		
Calepineae		
Arabideae	*Arabis paniculata*	**Zn, Cd**
Thlaspideae	*Peltaria emarginata*	**Ni**
Eutremeae		
Isatideae	*Isatis capadocica*	**As**
Brassiceae		
Schizopetaleae	*Streptanthus polygaloides*	**Ni**
	Stanleya pinnata	**Se**
Sysimbrieae		
Aphragmeae		
Biscutelleae		
Heliophileae		
Chorisporeae		
Malcomieae		
Hesperideae	*Hesperis persica*	**As**
Anchonieae		
Euclidieae		
Buniadeae		
Dontostemoneae		

◄ **Fig. 3.1** Phylogenetic tree of the Brassicaceae family, as reported by Lysak and Koch (2011). Currently accepted tribes are listed in the *first column*. Main hyperaccumulator species among the Brassicaceae are indicated in the *second column* together with the accumulated heavy metals, and are highlighted with *red squares*. *Dashed lines* indicate uncertain phylogenetic relationships. Branches are not drawn to scale. *Recent hypothesis for the subdivision of the Camelineae tribe, as suggested by Bailey et al. (2006)

diverged from its non-tolerant sister species *A. lyrata* around 337,000 years ago, with a speciation event coinciding with major adaptive changes that conferred hypertolerance (Roux et al. 2011). *A. halleri* is found mainly in Central and Eastern Europe, although the subspecies *gemmifera* occurs in Japan and Taiwan (Al-Shehbaz and O'Kane 2002). All *A. halleri* populations, from both non-contaminated and metalliferous soils, are constitutively able to hyperaccumulate Zn and Cd, although the degree of hyperaccumulation is variable and heritable (Macnair 2002; Meyer et al. 2010).

Noccaea caerulescens is a self-compatible diploid species, biannual or perennial, which shares an average sequence identity of 88% with *A. thaliana* within the coding regions (Assunção et al. 2003a; Rigola et al. 2006). Zn hypertolerance and accumulation is constitutive in this species, although the trait shows more variability than in *A. halleri* (Verbruggen et al. 2009; Plessl et al. 2010; Krämer 2010). Some *N. caerulescens* ecotypes can also accumulate Cd and Ni. Furthermore, variations in Cd hyperaccumulation among different ecotypes seem to correlate with different degrees of Zn accumulation (Assunção et al. 2003b; Roosens et al. 2003). In some populations from Southern France, Cd may even be necessary for optimal growth (Roosens et al. 2003). Like Cd, Ni hyperaccumulation in *N. caerulescens* appears to be non-constitutive and confined to some populations from serpentine soils (Assunção et al. 2003b). The most studied ecotypes are: Prayon (Belgium) and Ganges (France), both of which hyperaccumulate Zn and, in different degrees, Cd; Monte Prinzera (Italy) that accumulates Zn and Ni; La Calamine (Belgium), a Zn/Cd-hypertolerant population with low accumulation rates; and Lellingen (Luxembourg), a non-metalliferous population (Assunção et al. 2003a, b; Verbruggen et al. 2009).

3.2 Ecological Role of Metal Hyperaccumulation in Plants

Metal hyperaccumulation is an adaptive solution that may be disadvantageous for plants because it is associated with high energy costs and therefore slows metabolism and growth. However, the trait has evolved independently several times in different taxa, indicating that it must provide some evolutionary advantages. Several different explanations for metal hyperaccumulation have been proposed although in most cases there is no supporting experimental data. Six hypotheses were reviewed by Boyd and Martens (1992): metal tolerance/disposal, induction of drought resistance, interference, inadvertent uptake, and defense against herbivores and pathogens. Of these, the inadvertent uptake hypothesis gives no selective value to metal hyperaccumulators, regarding the trait as a by-product of other physiological processes.

According to the metal tolerance/disposal hypothesis, the accumulation of heavy metals in the aerial parts of the plant may contribute to tolerance by removing metals from sensitive tissues and eliminating them through the loss of leaves (Rascio and Navari-Izzo 2011). The drought resistance theory suggests that heavy metals could work as osmolytes in the cells. Neither hypothesis is supported by any experimental evidence.

The interference hypothesis takes allelopathy into consideration. Hyperaccumulators would be able to inhibit neighboring plants by creating a high metal-content zone, allowing them to compete with faster growing plants for space and light. However, most studies of elemental allelopathy have been inconclusive and do not take into consideration important criteria, such as the role hyperaccumulators play in increasing the metal content in the surrounding area (Morris et al. 2009). Exhaustive work comparing the Se hyperaccumulators *Atragalus bisulcatus* and *Stanleya pinnata* with the non-accumulators *Astragalus drummondii* and *Stanleya elata* growing in seleniferous and non-seleniferous soils demonstrated that plants can affect Se accumulation in their neighbors, and that Se in the soil influences competition and facilitation between plants. Therefore, Se hyperaccumulators may affect the composition of plant communities by allowing growth of Se-tolerant species (El Mehdawi et al. 2012).

Finally, the elemental defense hypothesis considers the role of heavy metals in defense against herbivores and pathogens, and is the most supported theory. The role of Ni (Jhee et al. 2006b), Cd (Jiang et al. 2005), Zn (Behmer et al. 2005), As (Rathinasabapathi et al. 2007), and Se (Galeas et al. 2008; Quinn et al. 2010) in protecting plants from biotic stresses has been confirmed. Biotic stress resistance is a direct effect of metal accumulation, since metal-tolerant pathogens show a greater ability to colonize hyperaccumulator plants (Fones et al. 2010). Defense is mediated both by the toxicity of heavy metals and their deterrent action, since herbivores seem to prefer plants that accumulate low levels of metals (Pollard and Baker 1997; Boyd et al. 2002). For example, many experiments conducted on *Brassica juncea* plants grown with or without Se, and exposed to caterpillars (*Pieris rapae*) and a fungal pathogen of the root system (*Fusarium* sp.), showed that caterpillars strongly preferred leaves without Se, and Se-containing plants were less susceptible to fungal infection (Hanson et al. 2003).

Elemental defense provides advantages over chemical defense because heavy metal uptake requires less metabolic effort than the biosynthesis of chemical toxins and deterrents, and inorganic elements cannot be degraded by herbivores, although some of them are able to chelate metals and therefore develop tolerance (Rascio and Navari-Izzo 2011). This is supported by the low levels of glucosinolate found in the metal hyperaccumulators *Streptanthus polygaloides* (Davis and Boyd 2000) and *N. caerulescens* (Tolrà et al. 2001). However, joint effects between different metals and between metals and chemical compounds have been demonstrated

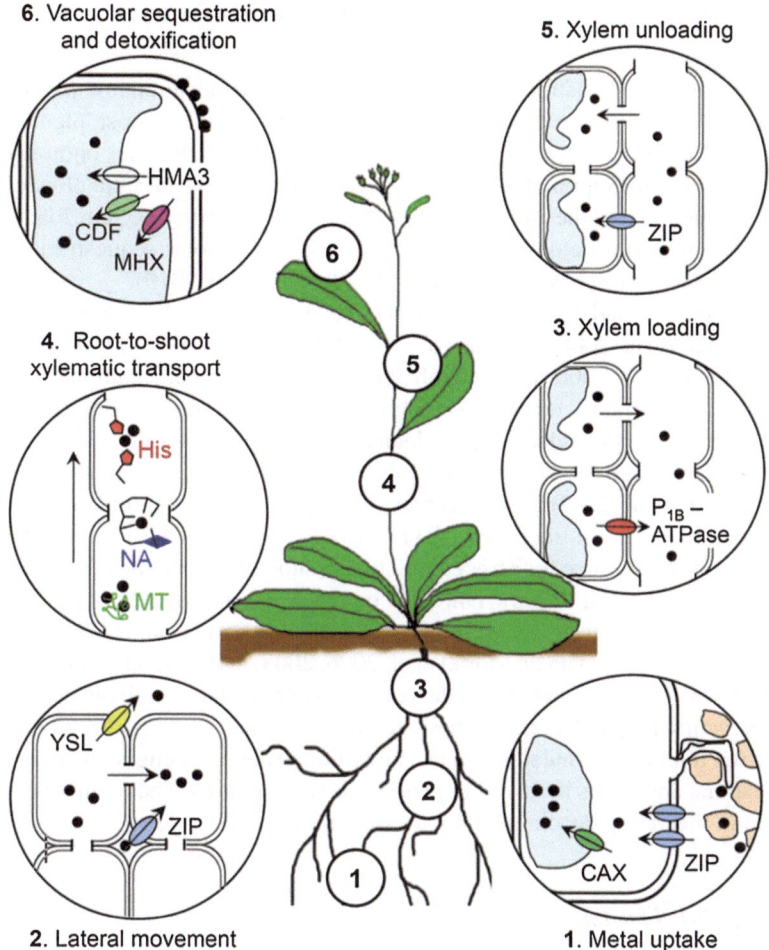

6. Vacuolar sequestration and detoxification

5. Xylem unloading

4. Root-to-shoot xylematic transport

3. Xylem loading

2. Lateral movement

1. Metal uptake

Fig. 3.2 Main mechanisms that are involved in metal accumulation by hyperaccumulating plants. The most relevant metal transporters and chelators described in this chapter are reported. The *black dots* represent metal ions and the *black arrows* indicate the direction of their transport. Areas inside the cells indicate the vacuole

(Jhee et al. 2006a). Recently, a proteomic approach aiming to unravel differences in the *A. halleri* proteome following treatment with Cd and Zn highlighted that proteins involved in plant defense mechanisms against biotic stress are down-regulated by heavy metals. In other words, if a high metal concentration in the shoot provides protection, then other defense mechanisms can be temporarily saved. These data also suggest there is cross-talk between heavy metal signaling and defense signaling (Farinati et al. 2009).

3.3 Determinants for Metal Hyperaccumulation

As stated above, the distinctive characteristic of hyperaccumulator plants is the partitioning of heavy metals in the aerial tissues, whereas most plant species confine metals to the roots, thus preventing damage to the photosynthetic machinery. Hyperaccumulation is achieved by enhancing certain physiological processes, such as uptake into the roots, symplastic mobility, xylem loading and unloading, and metal detoxification by chelation or vacuolar sequestration in the shoots (for a review see Verbruggen et al. 2009; Krämer 2010).

Hypertolerance and hyperaccumulation are quantitative characters (Bert et al. 2003; Assunção et al. 2006; Deniau et al. 2006; Filatov et al. 2006, 2007; Courbot et al. 2007; Willems et al. 2007). Although they coexist in hyperaccumulator species, segregation experiments using *inter*specific and *intra*specific crosses demonstrate that they are genetically independent (Macnair et al. 1999; Assunção et al. 2003c; Bert et al. 2003). At least three QTLs for Zn hyperaccumulation have been identified in *A. halleri* (Filatov et al. 2007) as well as one major QTL for Cd accumulation (Willems et al. 2010). In *N. caerulescens*, two QTLs have been identified for Zn and Cd accumulation in the roots, three for Zn accumulation in shoots, and one for Cd accumulation in shoots (Deniau et al. 2006). Different comparative approaches, including transcriptomic (Becher et al. 2004; Weber et al. 2004; Hammond et al. 2006; Rigola et al. 2006; Talke et al. 2006; van de Mortel et al. 2006, 2008) and proteomic analysis (Ingle et al. 2005b; Tuomainen et al. 2010), have been used to isolate candidate determinants of hyperaccumulation. Several genes are overexpressed in hyperaccumulators in comparison to non-accumulator species, including genes encoding metal transporters and chelators, and genes involved in generic stress responses. The main mechanisms involved in metal hyperaccumulation are summarized in Fig. 3.2.

3.3.1 Metal Transporters

3.3.1.1 ZIP Family

The ZIP family has been shown to promote cation (particularly Zn) uptake and accumulation in *A. thaliana* (Lin et al. 2009), suggesting that ZIP genes may play an important role in metal hyperaccumulation. Some ZIP genes have been isolated from hyperaccumulator species, including the *N. caerulescens* genes encoding NcZNT1 and NcZNT2, which are homologous to AtZIP4 (Assunção et al. 2001), as well as NcZNT5 and NcZNT6, which are homologous to AtZIP5 and AtZIP6, respectively (Wu et al. 2009). Mizuno et al. (2005) cloned the genes for TjZNT1, which transports Zn, Cd, and Mn, and TjZNT2, which is more specific for Zn and Mn, from the Ni hyperaccumulator *Thlaspi japonicum*. AhIRT3, which is involved in Fe and Zn transport, was identified in *A. halleri* (Lin et al. 2009), as well as CsZIP1 from the Mn hyperaccumulator *Chengiopanax sciadophylloides* (Mizuno et al. 2008).

Several ZIP genes do appear to be overexpressed in hyperaccumulator species, including *A. halleri* (*ZIP4* and *ZIP6*: Becher et al. 2004; *ZIP9*: Weber et al. 2004; *ZIP6*: Filatov et al. 2007; *IRT3*, *ZIP3*, *ZIP4*, *ZIP6*, *ZIP9*, and *ZIP10*: Talke et al. 2006) and *N. caerulescens* (*IRT3*, *ZIP6* and *ZIP7*: Hammond et al. 2006; *ZIP3*, *ZIP4* and *ZIP9*: van de Mortel et al. 2006; *ZIP1* and *ZIP8*: van de Mortel et al. 2008) in comparison to non-accumulators. The overexpression of ZIP genes may in some cases reflect gene duplication events, e.g., *AhZIP3*, *AhZIP6* and *AhZIP9* in *A. halleri* (Talke et al. 2006).

3.3.1.2 CDF Family

Members of the CDF family are important for the maintenance of metal homeostasis by mediating the efflux of metal ions from the cytosol into the apoplast or vacuole (Gustin et al. 2011). In particular, the *A. thaliana* CDF protein AtMTP1 induces Zn tolerance and accumulation when overexpressed in transgenic *A. thaliana* plants, suggesting a potential role in hyperaccumulation (van der Zaal et al. 1999).

Accordingly, AtMTP1 homologs in hyperaccumulators appear to have an important role. *MTP1* is strongly expressed in *A. halleri* (Becher et al. 2004; Dräger et al. 2004; Talke et al. 2006) and *N. caerulescens* (Assunção et al. 2001) in comparison to non-accumulator species, and cosegregates with the QTL for Zn tolerance in *A. halleri* (Willems et al. 2007; Shahzad et al. 2010). *MTP1* is also induced in the presence of Cd in *N. caerulescens*, suggesting a role in the response to Cd toxicity (Küpper and Kochian 2010).

A. halleri MTP1 genes have been studied in detail and five paralogs have been detected, named *AhMTP1-A1*, *-A2*, *-B*, *-C*, and *-D* (Dräger et al. 2004; Shahzad et al. 2010). *AhMTP1-D* is not fixed in at least one metalliculous population. The *AhMTP1* copies share on average 97.5% sequence identity, and respectively 91 and 93% identity with their orthologs in *A. thaliana* and *A. lyrata*. Stronger divergences are present in the promoter regions and are correlated with different expression levels in the different species (Shahzad et al. 2010). The *A. halleri* paralogs are differentially expressed and are modulated by Zn (Dräger et al. 2004; Shahzad et al. 2010). Exhaustive analysis of MTP1 has also been carried out in the hyperaccumulators *N. caerulescens* (Assunção et al. 2001) and *Thlaspi goesingense* (Kim et al. 2004; Gustin et al. 2009). In particular, the overexpression of TgMTP1 in *A. thaliana* induces a systematic response that includes the increased expression of Zn transporters (*ZIP3*, *ZIP4*, *ZIP5*, and *ZIP9*), suggesting that TgMTP1 may induce Zn accumulation by initiating a Zn deficiency response (Gustin et al. 2009).

AhMTP8 and *AhMTP11* are also overexpressed in *A. halleri* in comparison to *A. thaliana* (Talke et al. 2006). Similarly, *NcMTP8* is expressed more strongly in the presence of excess Zn in *N. caerulescens* in comparison to *A. thaliana* (van de Mortel et al. 2006). *NcMTP11* and *NcMTP12* also show higher expression levels in *N. caerulescens* in comparison to *T. arvense* (Hammond et al. 2006). These metal transporters belong to group I of the CDF family (Krämer et al. 2007) and are homologous to ShMTP8 (formerly ShMTP1) from the Mn-tolerant legume *Stylosanthes hamata*, which transports Mn into the vacuoles (Delhaize et al. 2003). NcMTP11 and NcMTP12 may therefore contribute to the homeostasis of metals other than Zn and Cd.

3.3.1.3 P_{1B}-Type ATPases

P_{1B}-type ATPases (HMAs) play a prominent role in the homeostasis of different metals. In *A. thaliana* they are involved both in Zn and Cd root-to-shoot translocation (AtHMA2 and AtHMA4: Wong and Cobbett 2009) and metal detoxification by vacuolar storage (AtHMA3: Morel et al. 2009), confirming their important role in heavy metal tolerance and accumulation.

HMA1 (Becher et al. 2004), *HMA3* (Becher et al. 2004; Filatov et al. 2006) and *HMA4* (Talke et al. 2006) are overexpressed in *A. halleri* in comparison to *A. thaliana*. *AhHMA4* was found to co-localize with the QTLs for Zn (Willems et al. 2007; Roosens et al. 2008) and Cd tolerance (Courbot et al. 2007) and for Zn and Cd accumulation (Willems et al. 2010). The downregulation of *AhHMA4* by RNA interference demonstrates that Zn and Cd tolerance and Zn hyperaccumulation are mainly due to AhHMA4, which seems to be responsible for loading metals into the xylem and their redistribution in the leaf blade (Hanikenne et al. 2008). Three almost identical gene copies are present in the *A. halleri* genome, suggesting a recent duplication event. In addition to the higher copy number, promoter modifications have enhanced the expression of *AhHMA4* (Hanikenne et al. 2008). The speciation event that separated *A. halleri* from its sister species *A. lyrata* may have coincided with the duplication of *HMA4* (Roux et al. 2011).

The P_{1B}-type ATPases *NcHMA3* and *NcHMA4* are overexpressed in *N. caerulescens* in comparison to *T. arvense* (Hammond et al. 2006) and *A. thaliana* (van de Mortel et al. 2006). *NcHMA4* is expressed more strongly in *N. caerulescens* roots than shoots and confers Cd resistance in yeast (Bernard et al. 2004). Four tandem copies of *NcHMA4* are present in the *N. caerulescens* genome. The paralogs share 88–99% sequence identity as well as 76–78% and 62–66% identity, respectively, with *A. thaliana* and *A. halleri*, indicating that the gene amplification is a relatively recent event within the *N. caerulescens* lineage (Ó Lochlainn et al. 2011). Each gene copy is constitutively expressed at high levels as in *A. halleri* (Ó Lochlainn et al. 2011). NcHMA3 is localized in the tonoplast and is highly specific for Cd. It is expressed at a higher level in the Ganges ecotype, which accumulates more Cd than Prayon. Overexpression of *NcHMA3* in *A. thaliana* induces Cd tolerance and low levels of Zn tolerance (Ueno et al. 2011).

3.3.1.4 NRAMP Family

NRAMP transporters are involved in the remobilization of metals (especially Fe) from the vacuole, which suggests a putative role in hyperaccumulation (Lanquar et al. 2005).

Accordingly, it was shown that *NRAMP3* is overexpressed in *A. halleri* (Weber et al. 2004; Filatov et al. 2006; Talke et al. 2006) and *N. caerulescens* (van de Mortel et al. 2006) in comparison to *A. thaliana*. NcNRAMP3 and NcNRAMP4 from *N. caerulescens* have been characterized and are similar in terms of localization and biological activity to their orthologs in *A. thaliana* (Oomen et al. 2009). NcNRAMP3 can transport Fe, Mn, and Cd in yeast, whereas NcNRAMP4 transports Zn in addition (Oomen et al. 2009). The expression of NcNRAMP3 in yeast also induces the accumulation of Cd but reduces Ni accumulation, underlining its role in heavy metal homeostasis (Wei et al. 2009). These data indicate that the differing roles of NRAMP proteins in *A. thaliana* and the hyperaccumulator *N. caerulescens* may reflect different levels or patterns of gene expression (Oomen et al. 2009). An NRAMP4 ortholog has also been cloned in the Ni-hyperaccumulator species *T. japonicum*, and the expression of TjNRAMP4 in yeast induces Ni accumulation and sensitivity, inferring a role in Ni homeostasis (Mizuno et al. 2005).

3.3.1.5 YSL Family

YSL transporters interact with a variety of heavy metals as phytosiderophore and NA chelates (Schaaf et al. 2004). Their involvement in the lateral translocation of metals into the veins (DiDonato et al. 2004) suggests a role in root-to-shoot translocation and therefore in metal hyperaccumulation. Nevertheless, few YSL genes appear to be overexpressed in hyperaccumulator species by transcriptomic approach, only *YSL6* in *A. halleri* (Talke et al. 2006) and *YSL7* in *N. caerulescens* (van de Mortel et al. 2006), suggesting that YSL proteins contribute minimally to the regulation of metals other than Fe. However, three *N. caerulescens* YSL genes (*NcYSL3*, *NcYSL5*, and *NcYSL7*) are expressed at higher levels than their orthologs in *A. thaliana* (Gendre et al. 2007). These genes are constitutively expressed at high levels around the vasculature, and they are not inducible by heavy metals. NcYSL3 can transport both Fe- and Ni–NA complexes in yeast assays (Gendre et al. 2007).

3.3.1.6 CaCA Superfamily

As discussed above, the MHX and CAX transporters are the only members of the CaCA family that appear to be involved in heavy metal accumulation in plants. MHX is a vacuolar Mg^{2+} and Zn^{2+}/H^+ antiport (Shaul et al. 1999). In *A. halleri*, the *AhMHX* gene is present as a single copy; the transcript is present mainly in the shoots and at similar levels to its ortholog in *A. thaliana*. However, the MHX protein is constitutive and much more abundant in *A. halleri* than in *A. thaliana*, indicating some form of post-transcriptional regulation (Elbaz et al. 2006).

Some CAX genes are overexpressed in metal hyperaccumulators, including *CAX2* in *A. halleri* (Becher et al. 2004; Weber et al. 2004), and *CAX2* (Hammod et al. 2006), *CAX3* (van de Mortel et al. 2008), and *CAX7* (van de Mortel et al. 2006) in *N. caerulescens*. Moreover, *NcCAX3* is induced by Cd (van de Mortel et al. 2008). These data suggest CAX genes are involved in metal hyperaccumulation. However, the metal specificity of CAX transporters has not been investigated thus far, although CAX2 does not transport Zn (Becher et al. 2004).

3.3.2 Metal Ligands

3.3.2.1 Histidine

Histidine is the most versatile free amino acid in terms of metal hyperaccumulation, and has a particularly high affinity for Ni (Callahan et al. 2006). The Ni-hyperaccumulator *Alyssum lesbiacum* accumulates high levels of histidine in the xylem sap when exposed to excess Ni (Krämer et al. 1996). Interestingly, Ni tolerance and Ni transport to shoots can also be induced in the non-accumulator species *Alyssum montanum* by feeding with histidine, underlining its important role in hyperaccumulation. Similarly, feeding the non-accumulator *Brassica juncea* with histidine increases Ni translocation by the xylem, although it has no impact on Ni uptake (Kerkeb and Krämer 2003).

The role of histidine biosynthesis in Ni accumulation was tested in *A. thaliana* by introducing the bacterial ATP phosphoribosyl transferase enzyme *StHisG*, which catalyzes the first step of histidine biosynthesis and is insensitive to feedback inhibition by histidine. The transgenic plants were much more Ni tolerant than wild type plants, confirming the important role of the free histidine pool (Wycisk et al. 2004). The relationship between histidine biosynthesis and Ni hyperaccumulation was also studied in the Ni hyperaccumulator *Alyssum lesbiacum*, by monitoring transcript and protein levels. This showed that there was no transcriptional regulation in response to excess Ni, but the levels of all enzymes (especially ATP phosphoribosyl transferase, ATP-PRT) were constitutively higher than those in the weak accumulator *Alyssum serpyllifolium* and the non-accumulator *Alyssum montanum* (Ingle et al. 2005a). The overexpression of AlATP-PRT from *Alyssum lesbiacum* in *A. thaliana* conferred Ni tolerance but had no impact on Ni accumulation (Ingle et al. 2005a) and there was no modulation of histidine biosynthesis by Ni in the Ni-hyperaccumulator *Thlaspi goesingense* (Persans et al. 1999). These data suggest that additional factors are necessary to develop a complete hyperaccumulation phenotype.

3.3.2.2 Nicotianamine

NA is the principal metal ligand in plants and it can form complexes with most transition metal ions (Verbruggen et al. 2009). A role for NA in Zn and Cd hyperaccumulation has been proposed because the genes in the NA biosynthesis pathway are upregulated in hyperaccumulators. *NAS2*, which is responsible for the last step in the pathway, is constitutively expressed in *A. halleri* roots at a higher level than its ortholog in *A. thaliana* (Becher et al. 2004; Weber et al. 2004). *NAS3* is overexpressed in shoots (Becher et al. 2004). *SAMS2*, which generates the NA precursor SAM, is also expressed at higher levels in *A. halleri* than *A. thaliana* (Talke et al. 2006). *N. caerulescens NAS3* and *NAS4* are also expressed at higher levels than their *A. thaliana* orthologs, and *NAS4* is constitutively expressed (van de Mortel et al. 2006, 2008). NA is also involved in Ni chelation, because Ni–NA complexes are found in *N. caerulescens* roots exposed to Ni (Vacchina et al. 2003). *NcNAS1* is constitutively expressed in shoots, whereas NA accumulation in roots appears to be Ni dependent (Mari et al. 2006). Furthermore, NcNAS1 overexpression in *A. thaliana* induces Ni tolerance and accumulation (Pianelli et al. 2005). Finally, NA is probably involved in the accumulation and mobilization of other metals, such as Fe, because *AhNAS4* cosegregates with a QTL for Fe accumulation in *A. halleri* (Willems et al. 2010).

3.3.2.3 Metallothioneins

Although no correlation with hyperaccumulation has been demonstrated, MTs are induced by several heavy metals e.g. in *A. thaliana* (Murphy and Taiz 1995) and are involved in metal tolerance and accumulation (Zimeri et al. 2005).

The *N. caerulescens MT2a* and *MT2b* genes are expressed at higher levels than their orthologs in *A. thaliana* (van de Mortel et al. 2006) and the *N. caerulescens MT2a* and *MT3* genes are expressed at higher levels than their orthologs in *T. arvense* (Hammond et al. 2006). The MT proteins were also expressed at higher levels in a metal-adapted *N. caerulescens* population in comparison to non-metalliculous populations (Hassinen et al. 2009). However, the different expression levels of *NcMT2a*, *NcMT2b*, and *NcMT3* do not correlate with the Cu, Cd, and Zn accumulation capacity and tolerance profiles in transgenic *A. thaliana* (Hassinen et al. 2009). There is no evidence of a direct connection between MTs as metal ligands and hyperaccumulation, thus it is likely that the increased tolerance induced by MT expression in some experiments is due to alternative roles, such as for ROS scavenging (Hassinen et al. 2011).

3.3.3 Response to Stress

3.3.3.1 Glutathione

Antioxidants are important for hyperaccumulators to address the potential oxidative stress caused by heavy metal ions. GS plays a key role for metal tolerance because it can act as a ROS scavenger, a metal chelator, and as a substrate for PC biosynthesis (Krämer 2010). The overexpression of genes involved in cysteine and GS biosynthesis has been reported in hyperaccumulators, e.g., the cysteine synthetase gene *OASA2* is expressed at higher levels in *A. halleri* than *A. thaliana* (Becher et al. 2004; Weber et al. 2004), and the glutathione-S-transferase protein GSTF10 is induced by heavy metals (Farinati et al. 2009). Similarly, *GSTF16* is overexpressed in *N. carulescens* shoots in comparison to *T. arvense* (Hammond et al. 2006), and other GST genes are expressed at higher levels than their *A. thaliana* orthologs (van de Mortel et al. 2008). Enhanced GS biosynthesis correlates with Ni tolerance, and the concentrations of GS, cysteine, and O-acetyl-L-serine (OAS) appear to correlate with Ni accumulation in different *Thlaspi* species, both hyperaccumulator and non-accumulator (Freeman et al. 2004). *A. thaliana* plants transformed with the enzyme serine acetyltransferase (SAT) from *T. goesingense* produce GS, cysteine, and OAS at similar levels to hyperaccumulator species and tolerate higher levels of Ni (Freeman et al. 2004), Zn, and Co, but not Cd (Freeman and Salt 2007). This probably reflects the fact that TgSAT is less sensitive to feedback inhibition induced by cysteine than AtSAT and can therefore accumulate higher levels of GS (Na and Salt 2011).

3.3.3.2 Lignin

The cell wall is an important site for metal storage in plants because it provides a large number of metal-binding sites (Maestri et al. 2010). Genes involved in phenylpropanoid (*PAL2*), lignin (CytP450 family), and suberin biosynthesis (*CER3*, *CER6* and some LTP genes) are overexpressed in *N. caerulescens* in comparison to *A. thaliana* (van de Mortel et al. 2006, 2008). Lignin and suberin deposition in *N. caerulescens* results in the lignification of endodermal cells, followed by the formation of a second layer of endodermis and the development of the Casparian strip, which does not occur in *A. thaliana* or *T. arvense*. Lignification and suberification may therefore help to prevent metal efflux from the vascular cylinder (van de Mortel et al. 2006, 2008).

3.3.3.3 Defensins

Metal hyperaccumulation also induces the expression of genes involved in stress-response signaling (biotic and abiotic) including defensins, which are strongly overexpressed in both *A. halleri* (Becher et al. 2004; Talke et al. 2006) and *N. caerulescens* (van de Mortel et al. 2006) in comparison to *A. thaliana*. Some defensins in *A. halleri* are modulated by Zn (Mirouze et al. 2006). The overexpression of AhPDF1.1 induces Zn tolerance in both yeast and *A. thaliana*. This can be explained by a Zn-chelation hypothesis, in which defensins (like MTs) use their cysteine-rich domains to bind metal ions. Alternatively, defensins may interfere with divalent cation transporters, reflecting their structural similarity to some channel-blocking peptides (Mirouze et al. 2006).

References

Al-Shehbaz IA, O'Kane SLJ (2002) Taxonomy and phylogeny of *Arabidopsis* (Brassicaceae). In: Somerville CR, Meyerowitz EM (eds) The *Arabidopsis* book. American Society of Plant Biologist, Rockville

Assunção AGL, Da Costa Martins P, De Folter S, Vooijs R, Schat H, Aarts MGM (2001) Elevated expression of metal transporter genes in three accessions of the metal hyperaccumulator *Thlaspi caerulescens*. Plant Cell Environ 24:217–226

Assunção AGL, Schat H, Aarts MGM (2003a) *Thlaspi caerulescens*, an attractive model species to study heavy metal hyperaccumulation in plants. New Phytol 159:351–360

Assunção AGL, Ten Bookum WM, Nelissen HJM, Vooijs R, Schat H, Ernst WHO (2003b) Differential metal-specific tolerance and accumulation patterns among *Thlaspi caerulescens* populations originating from different soil types. New Phytol 159:411–419

Assunção AGL, Ten Bookum WM, Nelissen HLM, Vooijs R, Schat H, Ernst WHO (2003c) A cosegregation analysis of zinc (Zn) accumulation and Zn tolerance in the Zn hyperaccumulator *Thlaspi caerulescens*. New Phytol 159:383–390

Assunção AGL, Pieper B, Vromans J, Lindhout P, Aarts MGM, Schat H (2006) Construction of a genetic linkage map of *Thlaspi caerulescens* and quantitative trait loci analysis of zinc accumulation. New Phytol 170:21–32

Bailey CD, Koch MA, Mayer M, Mummenhoff K, O' Kane SL Jr., Warwick SI, Windham MD, Al-Shehbaz IA (2006) Toward a global phylogeny of the Brassicaceae. Mol Biol Evol 23:2142–2160

Baker AJM, McGrath SP, Reeves RD, Smith JAC (2000) Chapter 5. metal hyperaccumulator plants: a review of the ecology and physiology of a biological resource for phytoremediation of metal-polluted soils. In: Terry N, Bañuelos G (eds) Phytoremediation of contamined soil and water. CRC Press LLC, Boca Raton

Becher M, Talke IN, Krall L, Krämer U (2004) Cross-species microarray transcript profiling reveals high constitutive expression of metal homeostasis genes in shoots of the zinc hyperaccumulator *Arabidopsis halleri*. Plant J 37:251–268

Behmer ST, Lloyd CM, Raubenheimer D, Stewart-Clark J, Knight J, Leighton RS, Harper FA, Smith JAC (2005) Metal hyperaccumulation in plants: mechanisms of defence against insect herbivores. Funct Ecol 19:55–66

Bernard C, Roosens N, Czernic P, Lebrun M, Verbruggen N (2004) A novel CPx-ATPase from the cadmium hyperaccumulator *Thlaspi caerulescens*. FEBS Lett 569:140–148

Bert V, Meerts P, Saumitou-Laprade P, Salis P, Gruber W, Verbruggen N (2003) Genetic basis of Cd tolerance and hyperaccumulation in *Arabidopsis halleri*. Plant Soil 249:9–18

Boyd RS, Martens SN (1992) The raison d'être for metal hyperaccumulation by plants. In: Baker AJM, Proctor J, Reeves RD (eds) The vegetation of ultramafic (serpentine) soils. Intercept Limited, Andover

Boyd RS, Davis MA, Wall MA, Balkwill K (2002) Nickel defends the South African hyperaccumulator *Senecio coronatus* (Asteraceae) against *Helix aspersa* (Mullusca: Pulmonidae). Chemoecology 12:91–97

Callahan DL, Baker AJM, Kolev SD, Wedd AG (2006) Metal ion ligands in hyperaccumulating plants. J Biol Inorg Chem 11:2–12

Clauss MJ, Koch MA (2006) Poorly known relatives of *Arabidopsis thaliana*. Trends Plant Sci 11:449–459

Courbot M, Willems G, Motte P, Arvidsson S, Roosens N, Saumitou-Laprade P, Verbruggen N (2007) A major quantitative trait locus for cadmium tolerance in *Arabidopsis halleri* co-localizes with HMA4, a gene encoding a heavy metal ATPase. Plant Physiol 144:1052–1065

Davis MA, Boyd RS (2000) Dynamics of Ni-based defence and organic defences in the Ni hyperaccumulator *Streptanthus polygaloides* Gray (Brassicaceae). New Phytol 146:211–217

de la Rosa G, Peralta-Videa JR, Montes M, Parsons JG, Cano-Aguilera I, Gardea-Torresdey JL (2004) Cadmium uptake and translocation in tumbleweed (*Salsola kali*), a potential Cd-hyperaccumulator desert plant species: ICP/OES and XAS studies. Chemosphere 55:1159–1168

Delhaize E, Kataoka T, Hebb DM, White RG, Ryan PR (2003) Genes encoding proteins of the cation diffusion facilitator family that confer manganese tolerance. Plant Cell 15:1131–1142

Deng DM, Shu WS, Zhang J, Zou HL, Lin Z, Ye ZH, Wong MH (2007) Zinc and cadmium accumulation and tolerance in populations of *Sedum alfredii*. Environ Pollut 147:381–386

Deniau AX, Pieper B, Ten Bookum WM, Lindhout P, Aarts MGM, Schat H (2006) QTL analysis of cadmium and zinc accumulation in the heavy metal hyperaccumulator *Thlaspi caerulescens*. Theor Appl Genet 113:907–920

DiDonato RJ Jr, Roberts LA, Sanderson T, Eisley RB, Walker EL (2004) *Arabidopsis Yellow Stripe-Like2* (*YSL2*): a metal-regulated gene encoding a plasma membrane transporter of nicotianamine–metal complexes. Plant J 39:403–414

Dräger DB, Desbrosses-Fonrouge AG, Krach C, Chardonnens AN, Meyer RC, Saumitou-Laprade P, Krämer U (2004) Two genes encoding *Arabidopsis halleri* MTP1 metal transport proteins co-segregate with zinc tolerance and account for high MTP1 transcript levels. Plant J 39:425–439

El Mehdawi AF, Cappa JJ, Fakra SC, Self J, Pilon-Smits EA (2012) Interaction of selenium hyperaccumulators and non accumulators during cocultivation on seleniferous or nonseleniferous soil—the importance of having good neighbors. New Phytol. doi:10.1111/j.1469-8137.2011.04043.x

Elbaz B, Shoshani-Knaani N, David-Assael O, Mizrachy-Dagri T, Mizrahi K, Saul H, Brook E, Berezin I, Shaul O (2006) High expression in leaves of the zinc hyperaccumulator *Arabidopsis halleri* of AhMHX, a homolog of an *Arabidopsis thaliana* vacuolar metal/proton exchanger. Plant Cell Environ 29:1179–1190

Farinati S, DalCorso G, Bona E, Corbella M, Lampis S, Cecconi D, Polati R, Berta G, Vallini G, Furini A (2009) Proteomic analysis of *Arabidopsis halleri* shoots in response to the heavy metals cadmium and zinc and rhizosphere microorganisms. Proteomics 9:4837–4850

Filatov V, Dowdle J, Smirnoff N, Ford-Lloyd B, Newbury HJ, Macnair MR (2006) Comparison of gene expression in segregating families identifies genes and genomic regions involved in a novel adaptation, zinc hyperaccumulation. Mol Ecol 15:3045–3059

Filatov V, Dowdle J, Smirnoff N, Ford-Lloyd B, Newbury HJ, Macnair MR (2007) A quantitative trait loci analysis of Zinc hyperaccumulation in *Arabidopsis halleri*. New Phytol 174:580–590

Fones H, Davis CAR, Rico A, Fang F, Smith JAC, Preston GM (2010) Metal hyperaccumulation armors plants against disease. PLoS Pathog 6:e1001093

Freeman JL, Salt DE (2007) The metal tolerance profile of *Thlaspi goesingense* is mimicked in *Arabidopsis thaliana* heterologously expressing serine acetyl-transferase. BMC Plant Biol 7:63–73

Freeman JL, Persans MW, Nieman K, Albrecht C, Peer W, Pickering IJ, Salt DE (2004) Increased glutathione biosynthesis plays a role in Nickel tolerance in *Thlaspi* Nickel hyperaccumulators. Plant Cell 16:2176–2191

Galeas ML, Klamper EM, Bennett LE, Freeman JL, Kondratieff BC, Quinn CF, Pilon-Smits EAH (2008) Selenium hyperaccumulation reduces plant arthropod loads in the field. New Phytol 177:715–724

Gendre D, Czernic P, Conéjéro G, Pianelli K, Briat JF, Lebrun M, Mari S (2007) TcYSL3, a member of the YSL gene family from the hyperaccumulator *Thlaspi caerulescens*, encodes a nicotianamine-Ni/Fe transporter. Plant J 49:1–15

Gustin JL, Loureiro ME, Kim D, Na G, Tikhonova M, Salt DE (2009) MTP1-dependent Zn sequestration into shoot vacuoles suggests dual roles in Zn tolerance and accumulation in Zn-hyperaccumulating plants. Plant J 57:1116–1127

Gustin JL, Zanis MJ, Salt DE (2011) Structure and evolution of the plant cation diffusion facilitator family of ion transporters. BMC Evol Biol 11:76–89

Hammond JP, Bowen HC, White PJ, Mills V, Pyke KA, Baker AJM, Whiting SN, May ST, Broadley MR (2006) A comparison of the *Thlaspi caerulescens* and *Thlaspi arvense* shoot transcriptomes. New Phytol 170:239–260

Hanikenne M, Talke IN, Haydon MJ, Lanz C, Nolte A, Motte P, Kroymann J, Weigel D, Krämer U (2008) Evolution of metal hyperaccumulation required *cis*-regulatory changes and triplication of *HMA4*. Nature 453:391–396

Hanson B, Garifullina GF, Lindblom SD, Wangeline A, Ackley A, Krames K, Norton AP, Lawrence CB, Pilon-Smits EAH (2003) Selenium accumulation protects *Brassica juncea* from invertebrate herbivory and fungal infection. New Phytol 159:461–469

Hassinen VH, Tuomainen M, Peräniemi S, Schat H, Kärenlampi SO, Tervahauta AI (2009) Metallothioneins 2 and 3 contribute to the metal-adapted phenotype but are not directly linked to Zn accumulation in the metal hyperaccumulator, *Thlaspi caerulescens*. J Exp Bot 60:187–196

Hassinen VH, Tervahauta AI, Schat H, Kärenlampi SO (2011) Plant metallothioneins—metal chelators with ROS scavenging activity? Plant Biol 13:225–232

Ingle RA, Mugford ST, Rees JD, Campbell MM, Smith JAC (2005a) Constitutively high expression of the Histidine biosynthetic pathway contributes to Nickel tolerance in hyperaccumulator plants. Plant Cell 17:2089–2106

Ingle RA, Smith JAC, Sweetlove LJ (2005b) Responses to nickel in the proteome of the hyperaccumulator plant *Alyssum lesbiacum*. Biometals 18:627–641

Jaffrè T, Brooks RR, Lee J, Reeves RD (1976) *Sebertia acuminata*: a hyperaccumulator of nickel from New Caledonia. Science 193:579–580

Jhee EM, Boyd RS, Eubanks MD (2006a) Effectiveness of metal–metal and metal–organic compounds combinations against *Plutella xylostella* (Lepidoptera: Plutellidae): implication for plant elemental defence. J Chem Ecol 32:239–259

Jhee EM, Boyd RS, Eubanks MD, Davis MA (2006b) Nickel hyperaccumulation by *Streptanthus polygaloides* protects against the folivore *Plutella xylostella* (Lepidoptera: Plutellidae). Plant Ecol 183:91–104

Jiang RF, Ma DY, Zhao FJ, McGrath SP (2005) Cadmium hyperaccumulation protects *Thlaspi caerulescens* from leaf feeding damage by thrips (*Frankliniella occidentalis*). New Phytol 167:805–814

Karimi N, Ghaderian SM, Raab A, Feldmann J, Meharg AA (2009) An arsenic-accumulating, hypertolerant Brassica, *Isatis cappadocica*. New Phytol 184:41–47

Kazantzis G (2000) Thallium in the environment and health effects. Environ Geochem Health 22:275–280

Kerkeb L, Krämer U (2003) The role of free Histidine in xylem loading of Nickel in *Alyssum lesbiacum* and *Brassica juncea*. Plant Physiol 131:716–724

Kim D, Gustin JL, Lahner B, Persans MW, Baek D, Yun DJ, Salt DE (2004) The plant CDF family member TgMTP1 from the Ni/Zn hyperaccumulator *Thlaspi goesingense* acts to enhance efflux of Zn at the plasma membrane when expressed in *Saccharomyces cerevisiae*. Plant J 39:237–251

Krämer U (2010) Metal hyperaccumulation in plants. Annu Rev Plant Biol 61:28.1–28.18

Krämer U, Cotter-Howells JD, Charnock JM, Baker AJM, Smith JAC (1996) Free histidine as a metal chelator in plants that accumulate nickel. Nature 379:635–638

Krämer U, Talke IN, Hanikenne M (2007) Transition metal transport. FEBS Lett 581:2263–2272

Küpper H, Kochian LV (2010) Transcriptional regulation of metal transport genes and mineral nutrition during acclimatization to cadmium and zinc in the Cd/Zn hyperaccumulator, *Thlaspi caerulescens* (Ganges population). New Phytol 185:114–129

Lanquar V, Lelièvre F, Bolte S, Hamès C, Alcon C, Neumann D, Vansuyt G, Curie C, Schröder A, Krämer U, Barbier-Brygoo H, Thomine S (2005) Mobilization of vacuolar iron by AtNRAMP3 and AtNRAMP4 is essential for seed germination on low iron. EMBO J 24:4041–4051

Lin YF, Liang HM, Yang SY, Boch A, Clemens S, Chen CC, Wu JF, Huang JL, Yeh KC (2009) *Arabidopsis* IRT3 is a zinc-regulated and plasma membrane localized zinc/iron transporter. New Phytol 182:392–404

Liu W, Shu WS, Lan CY (2004) *Viola baoshanensis*, a plant that hyperaccumulates cadmium. Chin Sci Bull 49:29–32

Liu X, Peng K, Wang A, Lian C, Shen Z (2009) Cadmium accumulation and distribution in populations of *Phytolacca americana* L. and the role of transpiration. Chemosphere 78:1136–1141

Lysak MA, Koch MA (2011) Phylogeny, Genome, and Karyotype Evolution of Crucifers (Brassicaceae). In: Schmidt R, Bancroft I (eds) Genetics and Genomics of the Brassicaceae. Springer-Verlag, Berlin

Macnair MR (2002) Within and between population genetic variation for zinc accumulation in *Arabidopsis halleri*. New Phytol 155:59–66

Macnair MR, Bert V, Huitson SB, Saumitou-Laprade P, Petit D (1999) Zinc tolerance and hyperaccumulation are genetically independent characters. Proc R Soc B 266:2175–2179

Maestri E, Marmiroli M, Visioli G, Marmiroli N (2010) Metal tolerance and hyperaccumulation: costs and trade-offs between traits and environment. Environ Exp Bot 68:1–13

Mari S, Gendre D, Pianelli K, Ouerdane L, Lobinski R, Briat JF, Lebrun M, Czernic P (2006) Root-to-shoot long-distance circulation of nicotianamine and nicotianamine–nickel chelates in the metal hyperaccumulator *Thlaspi caerulescens*. J Exp Bot 57:4111–4122

Meyer CL, Kostecka AA, Saumitou-Laprade P, Créach A, Castric V, Pauwels M, Frérot H (2010) Variability of zinc tolerance among and within populations of the pseudometallophyte species *Arabidopsis halleri* and possible role of directional selection. New Phytol 185:130–142

Minguzzi C, Vergnano O (1948) Il contenuto di nichel nelle ceneri di *Alyssum bertolonii*. Atti Soc Tosc Sc Nat 55:49–74

Mirouze M, Sels J, Richard O, Czernic P, Loubet S, Jacquier A, François IEJA, Cammue BPA, Lebrun M, Berthomieu P, Marquès L (2006) A putative novel role for plant defensins: a defensin from the zinc hyper-accumulating plant, *Arabidopsis halleri*, confers zinc tolerance. Plant J 47:329–342

Mizuno T, Usui K, Horie K, Nosaka S, Mizuno N, Obata H (2005) Cloning of three ZIP/Nramp transporter genes from a Ni hyperaccumulator plant *Thlaspi japonicum* and their Ni^{2+}-transport abilities. Plant Physiol Biochem 43:793–801

Mizuno T, Hirano K, Kato S, Obata H (2008) Cloning of ZIP family metal transporter genes from the manganese hyperaccumulator plant *Chengiopanax sciadophylloides*, and its metal transport and resistance abilities in yeast. Soil Sci Plant Nutr 54:86–94

Morel M, Crouzet J, Gravot A, Auroy P, Leonhardt N, Vavasseur A, Richaud P (2009) AtHMA3, a P1B-ATPase allowing Cd/Zn/Co/Pb vacuolar storage in *Arabidopsis*. Plant Physiol 149:894–904

Morris C, Grossl PR, Call CA (2009) Elemental allelopathy: processes, progress, and pitfalls. Plant Ecol 202:1–11

Murphy A, Taiz L (1995) Comparison of metallothionein gene expression and nonprotein thiols in ten *Arabidopsis* ecotypes. Plant Physiol 109:945–954

Na G, Salt DE (2011) Differential regulation of serine acetyltransferase is involved in nickel hyperaccumulation in *Thlaspi goesingense*. J Biol Chem 286:40423–40432

National Research Council, Committee on Biologic Effects of Atmospheric Pollutants (1974) Chromium. National Academies Press, Washington

National Research Council, Committee on Medical and Biological Effects of Environmental Pollutants (1977) Arsenic: medical and biological effects of environmental pollutants. National Academies Press, Washington

Ó Lochlainn S, Bowen HC, Fray RG, Hammond JP, King GJ, White PJ, Graham NS, Broadley MR (2011) Tandem Quadruplication of HMA4 in the Zinc (Zn) and Cadmium (Cd) Hyperaccumulator Noccaea caerulescens. PLoS One 6:e17814

Oomen RJFJ, Wu J, Lelièvre F, Blanchet S, Richaud P, Barbier-Brygoo H, Aarts MGM, Thomine S (2009) Functional characterization of NRAMP3 and NRAMP4 from the metal hyperaccumulator *Thlaspi caerulescens*. New Phytol 181:637–650

Patra M, Bhowmik N, Bandopadhyay B, Sharma A (2004) Comparison of mercury, lead and arsenic with respect to genotoxic effects on plant systems and the development of genetic tolerance. Environ Exp Bot 52:199–223

Persans MW, Yan X, Patnoe JMML, Krämer U, Salt DE (1999) Molecular dissection of the role of Histidine in Nickel hyperaccumulation in *Thlaspi goesingense* (Hálácsy). Plant Physiol 121:1117–1126

Pianelli K, Mari S, Marqués L, Lebrun M, Czernic P (2005) Nicotianamine over-accumulation confers resistance to nickel in *Arabidopsis thaliana*. Transgenic Res 14:739–748

Plessl M, Rigola D, Hassinen VH, Tervahauta A, Kärenlampi S, Schat H, Aarts MGM, Ernst D (2010) Comparison of two ecotypes of the metal hyperaccumulator *Thlaspi caerulescens* (J. & C. PRESL) at the transcriptional level. Protoplasma 239:81–93

Pollard AJ, Baker AJM (1997) Deterrence of herbivory by zinc hyperaccumulation in *Thlaspi caerulescens* (Brassicaceae). New Phytol 135:655–658

Quinn CF, Freeman JL, Reynolds RJB, Cappa JJ, Fakra SC, Marcus MA, Lindblom SD, Quinn EK, Bennett LE, Pilon-Smits EAH (2010) Selenium hyperaccumulation offers protection from cell disruptor herbivores. BMC Ecol 10:19–30

Rascio N, Navari-Izzo F (2011) Heavy metal hyperaccumulating plants: How and why do they do it? And what makes them so interesting? Plant Sci 180:169–181

Rathinasabapathi B, Rangasamy M, Froeba J, Cherry RH, McAuslane HJ, Capinera JL, Srivastava M, Ma LQ (2007) Arsenic hyperaccumulation in the Chinese brake fern (*Pteris vittata*) deters grasshopper (*Schistocerca americana*) herbivory. New Phytol 175:363–369

Reeves RD, Baker AJM (2000) Metal-accumulating plants. In: Raskin I, Ensley BD (eds) Phytoremediation of toxic metals—using plants to clean up the environment. Wiley, New York

Rigola D, Fiers M, Vurro E, Aarts MGM (2006) The heavy metal hyperaccumulator *Thlaspi caerulescens* expresses many species-specific genes, as identified by comparative expressed sequence tag analysis. New Phytol 170:753–766

Roosens N, Verbruggen N, Meerts P, Ximenez-Embun P, Smith JAC (2003) Natural variation in cadmium tolerance and its relationship to metal hyperaccumulation for seven populations of *Thlaspi caerulescens* from western Europe. Plant Cell Environ 26:1657–1672

Roosens NHCJ, Willems G, Godé C, Courseaux A, Saumitou-Laprade P (2008) The use of comparative genome analysis and syntenic relationships allows extrapolating the position of Zn tolerance QTL regions from *Arabidopsis halleri* into *Arabidopsis thaliana*. Plant Soil 306:105–116

Roux C, Castric V, Pauwels M, Wright SI, Saumitou-Laprade P, Vekemans X (2011) Does speciation between *Arabidopsis halleri* and *Arabidopsis lyrata* coincide with major changes in a molecular target of adaptation? PLoS One 6:e26872

Schaaf G, Ludewig U, Erenoglu BE, Mori S, Kitahara T, von Wirén N (2004) ZmYS1 functions as a proton-coupled symporter for phytosiderophore- and nicotianamine-chelated metals. J Biol Chem 279:9091–9096

Shahzad Z, Gosti F, Frérot H, Lacombe E, Roosens N, Saumitou-Laprade P, Berthomieu P (2010) The five AhMTP1 zinc transporters undergo different evolutionary fates towards adaptive evolution to zinc tolerance in *Arabidopsis halleri*. PLoS Genet 6:e1000911

Shaul O, Hilgemann DW, de Almeida Engler J, Van Montagu M, Inzé D, Galili G (1999) Cloning and characterization of a novel Mg2+/H+ exchanger. EMBO J 18:3973–3980

Sillanpää M, Jansson H (1992) Status of cadmium, lead, cobalt and selenium in soils and plants of thirty countries. Food Agric Org Soil bull 65:195

Talke IN, Hanikenne M, Krämer U (2006) Zinc-dependent global transcriptional control, transcriptional deregulation, and higher gene copy number for genes in metal homeostasis of the hyperaccumulator *Arabidopsis halleri*. Plant Physiol 142:148–167

Tolrà RO, Poschenrieder C, Alonso R, Barceló D, Barceló J (2001) Influence of zinc hyperaccumulation on glucosinolates in *Thlaspi caerulescens*. New Phytol 151:621–626

Tuomainen M, Tervahauta A, Hassinen V, Schat H, Koistinen KM, Lehesranta S, Rantalainen K, Häyrinen J, Auriola S, Anttonen M, Kärenlampi S (2010) Proteomics of *Thlaspi caerulescens* accessions and an interaccession cross segregating for zinc accumulation. J Exp Bot 61:1075–1087

Ueno D, Milner MJ, Yamaji N, Yokosho K, Koyama E, Zambrano MC, Kaskie M, Ebbs S, Kochian LV, Ma JF (2011) Elevated expression of TcHMA3 plays a key role in the extreme Cd tolerance in a Cd-hyperaccumulating ecotype of *Thlaspi caerulescens*. Plant J 66:852–862

Vacchina V, Mari S, Czernic P, Marquès L, Pianelli K, Schaumlöffel D, Lebrun M, Łobiński R (2003) Speciation of nickel in a hyperaccumulating plant by high-performance liquid chromatography–inductively coupled plasma mass spectrometry and electrospray MS/MS assisted by cloning using yeast complementation. Anal Chem 75:2740–2745

van de Mortel JE, Almar Villanueva L, Schat H, Kwekkeboom J, Coughlan S, Moerland PD, Ver Loren van Themaat E, Koornneef M, Aarts MGM (2006) Large expression differences in genes for Iron and Zinc homeostasis, stress response, and lignin biosynthesis distinguish roots of *Arabidopsis thaliana* and the related metal hyperaccumulator *Thlaspi caerulescens*. Plant Physiol 142:1127–1147

van de Mortel JE, Schat H, Moerland PD, Ver Loren van Themaat E, van der Ent S, Blankestijn H, Ghandilyan A, Tsiatsiani S, Aarts MGM (2008) Expression differences for genes involved in lignin, glutathione and sulphate metabolism in response to cadmium in *Arabidopsis thaliana* and the related Zn/Cd-hyperaccumulator *Thlaspi caerulescens*. Plant Cell Environ 31:301–324

van der Zaal BJ, Neuteboom LW, Pinas JE, Chardonnens AN, Schat H, Verkleij JAC, Hooykaas PJJ (1999) Overexpression of a novel *Arabidopsis* gene related to putative zinc-transporter genes from animals can lead to enhanced zinc resistance and accumulation. Plant Physiol 119:1047–1055

Verbruggen N, Hermans C, Schat H (2009) Molecular mechanisms of metal hyperaccumulation in plants. New Phytol 181:759–776

Wang J, Zhao FJ, Meharg AA, Raab A, Feldmann J, McGrath SP (2002) Mechanisms of Arsenic hyperaccumulation in *Pteris vittata*. Uptake kinetics, interactions with phosphate, and arsenic speciation. Plant Physiol 130:1552–1561

Weber M, Harada E, Vess C, Von Roepenack-Lahaye E, Clemens S (2004) Comparative microarray analysis of Arabidopsis thaliana and Arabidopsis halleri roots identifies nicotianamine synthase, a ZIP transporter and other genes as potential metal hyperaccumulation factors. Plant J 37:269–281

Wei W, Chai T, Zhang Y, Han L, Xu J, Guan Z (2009) The *Thlaspi caerulescens* NRAMP homologue TcNRAMP3 is capable of divalent cation transport. Mol Biotechnol 41:15–21

Willems G, Dräger DB, Courbot M, Godé C, Verbruggen N, Saumitou-Laprade P (2007) The genetic basis of zinc tolerance in the metallophyte *Arabidopsis halleri* ssp. *halleri* (Brassicaceae): an analysis of quantitative trait *loci*. Genetics 176:659–674

Willems G, Frérot H, Gennen J, Salis P, Saumitou-Laprade P, Verbruggen N (2010) Quantitative trait loci analysis of mineral element concentrations in an *Arabidopsis halleri* x *Arabidopsis lyrata petraea* F$_2$ progeny grown on cadmium-contaminated soil. New Phytol 187:368–379

Wong CKE, Cobbett CS (2009) HMA P-type ATPases are the major mechanism for root-to-shoot Cd translocation in *Arabidopsis thaliana*. New Phytol 181:71–78

Wu J, Zhao FJ, Ghandilyan A, Logoteta B, Olortegui Guzman M, Schat H, Wang X, Aarts MGM (2009) Identification and functional analysis of two ZIP metal transporters of the hyperaccumulator *Thlaspi caerulescens*. Plant Soil 325:79–95

Wycisk K, Kim EJ, Schroeder JI, Krämer U (2004) Enhancing the first enzymatic step in the histidine biosynthesis pathway increases the free histidine pool and nickel tolerance in *Arabidopsis thaliana*. FEBS Lett 578:128–134

Zhao FJ, Dunham SJ, McGrath SP (2002) Arsenic hyperaccumulation by different fern species. New Phytol 156:27–31

Zimeri AM, Dhankher OP, McCaig B, Meagher RB (2005) The plant MT1 metallothioneins are stabilized by binding cadmium and are required for cadmium tolerance and accumulation. Plant Mol Biol 58:839–855

Chapter 4
Phytoremediation: The Utilization of Plants to Reclaim Polluted Sites

Andrea Nesler and Antonella Furini

Abstract The contamination of water and soil with heavy metals is a serious environmental and health hazard. Phytoremediation is the use of plants to remove pollutants, and this offers an alternative to conventional clean-up methods which rely on excavation or the application of detergents and other chemicals. The ideal plant for phytoremediation should tolerate heavy metals and accumulate them in the aerial tissues, should produce large amounts of biomass rapidly, and should develop a deep and extensive root system. Biotechnology offers the opportunity to genetically engineer plants that tolerate and accumulate large amounts of heavy metals in their shoots or that chemically transform and volatilize them. The uptake of heavy metals into plants can also be enhanced by the microbial community in the rhizosphere, which can stimulate root proliferation and increase metal bioavailability.

Keywords Phytoremediation · Transgenic plants · Metal tolerance · Effects of rhizosphere microbes

4.1 A Green Technology to Remove Heavy Metals from Soil and Water

Agriculture, mining, metallurgy, fossil fuel use, and military operations have dispersed large quantities of heavy metals and metalloids into the environment, posing a serious risk to biodiversity and human health (Wernick and Themelis 1998; Wijnhoven et al. 2007). Conventional methods for cleaning up contaminated sites

A. Nesler · A. Furini (✉)
Department of Biotechnology, University of Verona,
Strada Le Grazie 15, 37134 Verona, Italy
e-mail: antonella.furini@univr.it

A. Nesler
e-mail: andrea.nesler@univr.it

A. Furini (ed.), *Plants and Heavy Metals*, SpringerBriefs in Biometals,
DOI: 10.1007/978-94-007-4441-7_4, © Nesler and Furini 2012

include washing, excavation, and reburial of solid matrices such as soil, and pumping and treating systems for contaminated water (Glass 1999). These approaches are expensive, generally inefficient if the contaminant is present at a low concentration, and often causes significant changes to the physicochemical characteristics of soils.

A different method to reclaim polluted sites is to use plants that reduce the toxicity of metals in soil or water, a concept known as *phytoremediation* (Doty 2008; Eapen and D'Souza 2005; Cherian and Oliveira 2005). This is advantageous because plants are autotrophic organisms that use carbon dioxide and sunlight as sources of carbon and energy, and the decontamination can be carried out in situ, thus minimizing costs and human exposition (Salt et al. 1998). Phytoremediation is esthetically pleasing and generally acceptable to the public as it is considered a clean and green alternative to conventional approaches. Other advantages include the low installation and maintenance costs, the low environmental impact (particularly on soil properties), the use of harvested plant biomass to produce biofuel, and the ability to avoid spreading pollutants by wind and water. Nevertheless, phytoremediation has also disadvantages, including the slow rate of contaminant removal, inadequacy to deal with heavily polluted sites, and the risk that toxic compounds are made more bioavailable. Further limiting factors may include soil properties, the extent of the contamination, and climate conditions, so these must be considered during the development of phytoremediation programs (Eapen and D'Souza 2005; Pilon-Smits 2005).

The use of plants to decontaminate polluted areas can be implemented using several different strategies: (i) *phytoextraction*, in which plants take up metals and accumulate them in harvestable tissues aboveground; (ii) *rhizofiltration*, which is mainly concerned with the remediation of contaminated water, in which metals are either adsorbed at the root surface or absorbed by the roots; (iii) *phytostabilization*, in which plant metabolism is used to chemically stabilize the metal in the soil matrix, limiting its migration, bioavailability, and hazard; and (iv) *phytovolatilization*, in which plants take up contaminants from the substrate and release them into the atmosphere via the transpiration stream (Doty 2008; Eapen and D'Souza 2005; Cherian and Oliveira 2005). All these phytoremediation strategies have been used successfully. For example *Brassica juncea* plants have been used to reduce Pb contamination in soil by phytoextraction (Blaylock 2000), and sunflower plants have been used to remove U from contaminated water by rhizofiltration (Dushenkov and Kapulnik 2000).

4.2 Plants Suitable for the Phytoremediation of Heavy Metals

The efficiency of phytoremediation depends on the plant species, the concentration and combination of toxic metals, and (as a consequence of the above) on the agronomic practices that need to be optimized, such as the addition of fertilizers or chelators and pH adjustment. Researchers initially focused on hyperaccumulator plants, which possess the mandatory characteristics for phytoremediation, i.e., metal tolerance and accumulation. However, most hyperaccumulators grow slowly, produce limited biomass, have shallow roots, and tend to prefer specific habitats, all of which argue against their use (Chaney et al. 2005). Ideal plants for phytoremediation should (i) possess an intrinsic capacity to tolerate metals and to concentrate them in harvestable aerial tissues or to adsorb it on the root surface; (ii) grow rapidly and produce lots of biomass; (iii) develop a deep and extensive root system; (iv) be widely distributed, allowing growth in many habitats; and (v) should be easy to cultivate and harvest.

Because natural hyperaccumulators lack these necessary properties, a different strategy is to use conventional breeding and/or genetic engineering to create plants suitable for phytoremediation. Metal uptake, transport, accumulation, and detoxification mechanisms have been studied extensively in model species and in metal hyperaccumulators. Many of the genes that control these processes have been cloned and characterized. These genes can be introduced into any genetically amenable species that is suitable for bioremediation, including *Brassica juncea*, sunflower (*Helianthus annuus*), yellow poplar (*Liliodendron tulipifera*), and shrub tobacco (*Nicotiana glaucum*) (Eapen and D'Souza 2005).

4.3 Engineering Plant Metal Tolerance and Accumulation

Metal tolerance and accumulation can be achieved by efficient sequestration of metals into subcellular compartments, by overexpressing molecules involved in metal chelation and finally by increasing the expression of stress response genes. The processes influencing heavy metal accumulation in plants, and consequently the targets for genetic modification are: (i) the mobilization and uptake of metals from the soil; (ii) the formation of metal complexes and their transfer into the vacuoles of root cells for detoxification; (iii) the translocation of metal ions to shoots via the symplast or apopolast; (iv) the distribution of metals in aerial tissues; (v) the sequestration of metals within green tissues; and (vi) the translocation of accumulated metals and metalloids to less metabolically active cells such as trichomes (Clemens et al. 2002). Therefore, one way to enhance the phytoremediation capacity of plants is to target mechanisms responsible for metal tolerance: overexpressing rate limiting enzymes, increasing the production of different metal transporter proteins and enzymes involved in the repair of metal-induced damage, and operating on the signal transduction pathway by affecting the expression of regulatory proteins. The latter, in particular, has been exploited to increase Cd tolerance and accumulation in *A. thaliana* and tobacco. A transcription factor that is upregulated by Cd was isolated in *Brassica juncea* (BjCdR15; Fusco et al. 2005), and identified as the ortholog of *A. thaliana* TGA3. BjCdR15 is a regulator of Cd uptake, transport, and accumulation in the shoot. The constitutive overexpression of BjCdR15 in *A. thaliana* and tobacco increased Cd accumulation in the shoots, and the transgenic plants produced more shoot biomass and suffered less chlorosis than control plants. BjCdR15 also induced the expression of genes responsible for root-to-shoot Cd transport and implicated in the extrusion of Cd into the apoplast and vacuole (Farinati et al. 2010).

The metabolism of metal ions has also been enhanced by somatic hybridization. Protoplast fusion between the high-biomass species *Brassica napus* and the hyperaccumulator *Thlaspi caerulescens* was followed by the regeneration of hybrids that accumulate levels of Zn and Cd that would be toxic to *B. napus* (Brewer et al. 1999). An alternative approach is genetic transformation with *Agrobacterium rhizogenes*, which results in a fast-growing root culture with a more extensive root surface. *T. caerulescens* hairy root cultures tolerated high levels of Cd and accumulated large amounts of the metal (Nedelkoska and Doran 2000).

4.3.1 Transgenic Plants Overexpressing Metal Transporter Proteins

The transport of metal and metalloid cations across the plasma membrane and internal organellar membranes is an important checkpoint in the metal homeostasis network. The expression of metal transporters in transgenic plants is therefore a promising phytoremediation strategy (Krämer et al. 2007). The tissue distribution and intracellular location of a transporter protein (e.g., plasma membrane, tonoplast) determines its role. Therefore overexpression may increase metal uptake, translocation, and/or sequestration. Some examples of metal transporters which have been used to create transgenic plants that accumulate heavy metals are discussed below.

The tobacco calmodulin-binding protein NtCBP4 is a membrane-bound transporter related to the *A. thaliana* non-selective cation channel AtCNGC1 (74.2% identity; Kohler et al. 1999) and the *Hordeum vulgare* protein HvCBT1 (61.2% identity; Schuurink et al. 1998). Transgenic tobacco lines overexpressing NtCBP4 were tested with different metal ions, and were shown to be more Ni tolerant than wild-type plants, reflecting the slower intracellular accumulation of Ni. However, the transgenic plants were more sensitive to Pb, which was taken up and translocated to the shoots more efficiently and accumulated at levels 30% higher than wild-type plants. NtCBP4 was the first protein related to Pb accumulation identified in plants (Arazi et al. 1999).

Overexpression of the *A. thaliana* vacuolar metal ion/H^+ antiporters CAX2 and CAX4 in tobacco increased the ability of the transgenic plants to detoxify Cd^{2+}, Zn^{2+}, and Mn^{2+}, and increased the accumulation of Cd in root vacuoles (Korenkov et al. 2007). Similarly, transgenic plants expressing AtMTP were more resistant to Zn, and accumulated large amounts of the metal in their roots when excess Zn was present in the environment (van der Zaal et al. 1999).

The yeast transporter protein YCF1, a member of the ABC transporter family, pumps GS-Cd complexes into the vacuole (Li et al. 1997). Because Pb can also form complexes with thiol groups, transgenic *A. thaliana* plants overexpressing YCF1 were tested for Cd and Pb tolerance and accumulation, revealing higher levels of Cd in the vacuoles and overall higher levels of Cd and Pb (Song et al. 2003).

HMA4 encodes a plasma membrane protein of the 1_B family of transition metal pumps in the P-type ATPase superfamily (Courbot et al. 2007). The ectopic overexpression of AtHMA4 in *A. thaliana* increased Zn and Cd accumulation in the shoots (Courbot et al. 2007). This gene is also expressed at a higher level in the hyperaccumulator *A. halleri* compared to *A. thaliana* (Talke et al. 2006), which reflects the combination of modified *cis*-regulatory sequences and a higher gene copy number (Hanikenne et al. 2008). The distribution of Zn was changed in the roots of transgenic *A. thaliana* plants expressing AhHMA4 cDNA under the control of its native promoter, and the Zn deficiency response genes ZIP4 and IRT3 were also upregulated in these plants, suggesting Zn partitioning into xylem vessels and consequent upregulation of Zn deficiency response genes.

The MerC plasma membrane transporter that controls Hg^{2+} uptake and resistance in bacteria was expressed in *A. thaliana* to improve Hg^{2+} accumulation. The rate of Hg^{2+} accumulation in detached leaves was higher in the transgenic plants than wild-type plants, although the transgenic seedlings were hypersensitive to Hg^{2+} (Sasaki et al. 2006). The possibility of altering the metal specificity was also tested. The *A. thaliana* iron transporter IRT1 can transport Fe, Zn, Mn, and Cd. The substitution of conserved residues capable of binding metal to alanine resulted in either altered substrate specificity or the abolition of transport (Rogers et al. 2000).

These studies have helped to determine the molecular mechanisms underlying heavy metal transport and accumulation in plants, showing that the overexpression of metal transport genes either identified in plants or in other species, can enhance metal transport and accumulation/sequestration in plant cells. These studies will contribute to the development of phytoremediation technologies that can be applied to soil polluted with heavy metals and indicate that it is also possible to design transgenic plants that accumulate specific metals.

4.3.2 Focusing on Metal Chemistry in Plants

4.3.2.1 Mercury as a Case Study

Plants do not volatilize Hg to a significant extent but accumulate it in roots and shoots. The use of plants for Hg phytoextraction therefore presents some limitations, since they cannot detoxify methylmercury and their low Hg tolerance limits their suitability for phytoremediation. The introduction in plant of a bacterial pathway that converts methylmercury to volatile elemental Hg (Hg^0) allowed overcoming the mentioned limitations. The *merB* gene encodes an *organomercurial lyase* that converts methylmercury to Hg^{2+}, whereas *merA* encodes a *mercuric reductase* and reduces Hg^{2+} to elemental Hg, using NADPH as the electron donor. To use these genes for phytoremediation, they were modified to optimize their expression in plants and placed under the control of the constitutive CaMV 35S promoter. Transgenic *A. thaliana* plants expressing *merA* tolerated high levels of Hg^{2+} and were able to volatize elemental Hg (Rugh et al. 1996), whereas those expressing *merB* were able to tolerate 10-fold higher levels of methylmercury and other organomercurials than untransformed plants (Bizily et al. 2000). Double transgenic plants (*MerA-MerB*) were able to tolerate 50-fold higher levels of organomercurials than wild-type plants and (unlike the single transgenics) were able to volatilize elemental Hg when supplied with organic Hg. These promising results indicate that the same gene constructs could be used to achieve Hg volatilization in other species. The phytovolatilization approach has been criticized because it releases volatile Hg into the atmosphere, but the wide dispersion and extensive dilution are thought to overcome any potential risks (Moreno et al. 2005).

4.3.2.2 Arsenic as a Case Study

In soil and water, As exists primarily in its oxidized form as the oxyanion arsenate (AsO_4^{3-}). As discussed above, the chemical similarity between arsenate and phosphate allows arsenate to be taken up from the soil via the phosphate pathway (Meharg et al. 1992). Arsenite (AsO_3^{3-}), the reduced form of As, has a very strong affinity for thiol groups, and can be sequestered into peptide-thiol complexes in plant shoots. Therefore, transgenic plants were tailored to accumulate more As by controlling the oxidation state of As and increasing the presence of thiol groups (Dhankher et al. 2002). This has been achieved by expressing two bacterial genes in *A. thaliana*, an arsenate reductase (*arsC*) and a γ–*ECS*. Double transgenic plants (ArsC-γ–ECS) were able to tolerate higher levels of arsenate and were able to accumulate glutathione-arsenite complexes in the shoots when grown in arsenate-containing hydroponic medium. The transgenic plants grew well in the presence of 200 μM arsenate, suggesting that other plant species could be transformed with *ArsC* and γ–*ECS* to allow the accumulation of As in aerial tissues when growing on As-contaminated soil (Dhankher et al. 2002).

4.3.2.3 Selenium as a Case Study

Preliminary in vitro studies suggested that the assimilation of selenate by plants is limited because the compound is reduced by APS (Shaw and Anderson 1972). Plants supplemented with selenate mostly accumulate selenate, whereas plants supplemented with selenite mostly accumulate organic Se (de Souza et al. 1998; Zayed et al. 1998). *B. juncea* plants overexpressing *A. thaliana* APS1 accumulate organic Se in the presence of selenate, whereas wild-type plants accumulate selenate (Pilon-Smits et al. 1999). This study demonstrated that APS1 reduces selenate but is rate limiting for selenate uptake and assimilation. The toxicity of Se predominantly reflects the incorporation of selenocysteine into proteins, disrupting protein structure and function. The conversion of selenocysteine into less toxic forms of Se is a tolerance mechanism adopted by Se hyperaccumulators such as *Astragalus bisulcatus*. SMT methylates selenocysteine to produce the non-protein amino acid methylselenocysteine, thus reducing the intracellular concentration of selenocysteine and selenomethionine and inhibiting the incorporation of these toxic forms of Se into proteins. The overexpression of *A. bisulcatus* SMT in *A. thaliana* and *B. juncea* caused the plants to accumulate more methylseleno-cysteine than wild-type plants, and to increase Se accumulation and volatilization (Le Duc et al. 2004). The effect of SMT in the transgenic *B. juncea* plants was more substantial when the plants were exposed to selenite rather than to selenate, indicating that the conversion of selenate to selenite is the limiting step in selenocysteine biosynthesis, reducing the potential benefit of SMT overexpression. To overcome this limitation, double transgenic plants were produced (by crossing SMT and APS transgenic lines) and these produced eight times more methylselenocysteine than wild-type plants and twice the amount of SMT transgenic plants (LeDuc et al. 2006). The double transgenic plants also accumulated nine times more total Se than wild-type plants (LeDuc et al. 2006). The combined expression of these two enzymes enhanced selenate uptake, conversion to selenite and selenocysteine methylation, therefore increasing the amount of Se uptake but detoxifying it at the same time. A further step toward Se phytoremediation is the conversion of selenocysteine to volatile dimethylsel-enide because the volatilization of Se from selenomethionine is much faster than from selenocysteine. CgS is the first enzyme in the selenocysteine to selenome-thionine pathway and transgenic *B. juncea* plants expressing *A. thaliana* CgS achieved a higher volatilization rate than wild-type plants (van Huysen et al. 2003). These plants also accumulated 40% less Se than wild-type plants, probably reflecting the enhanced volatilization (Pilon-Smits and LeDuc 2009).

4.3.3 Transgenic Plants Overexpressing Metal Chelating Compounds

The overproduction of several metal chelators has demonstrated to affect plant metal tolerance and accumulation, and different research groups have tested recombinant metal binding proteins. In many cases tolerance is achieved by overexpressing MT without metal ions accumulating to higher levels than normal. For example, the expression of CUP1 in tobacco promoted the accumulation of Cu but not Cd in the leaves (Thomas et al. 2003). Similarly, the overexpression of a pea MT in *A. thaliana* induced the accumulation of Cu (Evans et al. 1992). Recently, an optimized mouse *mt1* gene was expressed in the tobacco chloroplast genome to promote Hg accumulation, providing an ideal strategy for Hg phytoremediation (Ruiz et al. 2011).

Since PCs are necessary for Cd tolerance it was reasonable that the overexpression of PCS would result in overproduction of these metal binding peptides and hence in higher Cd tolerance and accumulation and it was also considered that such transgenic plants could be used to phytoremediate Cd-contaminated sites (Wojas et al. 2008). Unfortunately, experimental results have been contradictory. The expression of *A. thaliana PCS1* in *E. coli* (Sauge-Merle et al. 2003) and yeast (Vatamaniuk et al. 1999) induced the accumulation of Cd, suggesting that PCs are involved in chelation or sequestration. However, PCS1 overexpression in *A. thaliana* enhanced PC biosynthesis but induced Cd hypersensitivity. The expression of wheat PCS in tobacco increased Cd and Pb accumulation (Martinez et al. 2006), whereas Peterson and Oliver (2006) showed that the overexpression of *AtPCS1* in *A. thaliana* leaves increased Cd tolerance without influencing the metal content of the shoots. Long-distance root-to-shoot transport of PC-Cd was observed in *A. thaliana* plants expressing wheat PCS, resulting in the reduction of Cd levels in roots (Gong et al. 2003). Pomponi et al. (2006) reported that tobacco plants expressing *A. thaliana* PCS were more Cd tolerant although Cd accumulated only when the medium was supplemented with GSH. Transgenic *B. juncea* plants overexpressing γ–ECS or GSS showed higher Cd tolerance and accumulation (Zhu et al. 1999a, b). Despite the abundant literature regarding the role of PCs in heavy metal detoxification and accumulation in plants, it is impossible to assign a clear role of these binding peptides in metal accumulation and hence in phytoremediation.

4.4 Effects of Rhizosphere Microbes on Phytoremediation

The absorption of metals by plants is strongly influenced by microbes in the rhizosphere. Carbon compounds released into the rhizosphere by plants provide energy for fungi and bacteria. Therefore the microbial population in the rhizosphere is much richer than in the bulk soil. It has been reported that Se and Hg uptake by plants can also be enhanced by rhizosphere bacteria (De Souza et al. 1999) and that plant metal uptake can as well be stimulated by mycorrhizal fungi (Frey et al. 2000) reflecting the stimulation of root proliferation and/or a microbial influence on metal bioavailability (De Souza et al. 2000). There is a positive correlation between rhizospheric microbes and Zn/Cd accumulation in *A. halleri* shoots (Farinati et al. 2009). However, when *A. halleri* plants were cultivated in the presence of few bacterial strains, selected from the rhizosphere population for Zn and Cd resistance, the accumulation of these metals in the shoots was much lower, suggesting that the presence of these bacterial strains impaired metal uptake (Farinati et al. 2011). Some of the strains affected the shoot metal content more than others, highlighting the important role of microbes in the accumulation of metals by plants and emphasizing that successful phytoremediation may depend on the selection of an appropriate rhizobacterial consortium.

Phytoextraction could be enhanced by supplementing the soil with bacteria that promote plant growth by stimulating germination and biomass production during phytoremediation. Such bacteria can affect plant health *indirectly*, by inhibiting the growth of pathogenic soil microbes, and *directly* by providing plants with growth-promoting compounds or by facilitating the uptake of nutrients from soil (Glick 2003). Many studies has been focused on metal hyperaccumulator species investigating the molecular basis of uptake, transport, and accumulation (Verbruggen et al. 2009) but further research is needed to investigate the interactions between hyperaccumulators and microbes where these communities are adapted.

4.5 Perspectives

Phytoremediation has received considerable attention from researchers and is widely viewed as a green technology for the remediation of polluted environments, but it is not yet exploited as a commercial platform technology. However, the increasing pace of genome sequencing will lead to the discovery of further metal-related genes that can be tested in transgenic plants for phytoremediation applications. Gene transfer techniques already allow plastid targeting and the simultaneous transfer and expression of several genes. Therefore, novel information about plant metabolism and plant–microbe interactions will facilitate the development of novel phytoremediation strategies to clean up soils polluted with heavy metals.

If the phytoremediation of polluted soils is possible using transgenic plants that can accumulate metals without negative effects, then it should also be possible to enhance crops with micronutrients such as Fe and Zn, particularly in areas where soils are depleted or where bioavailability is limited. A better understanding of the genetic factors that control the accumulation, metabolism, and tolerance of metals in plants will allow plants to be utilized efficiently for both phytoremediation and biofortification.

References

Arazi T, Sunkar R, Kaplan B, Fromm H (1999) A tobacco plasma membrane calmodulin-binding transporter confers. Ni^{2+} tolerance and Pb^{2+} hypersensitivity in transgenic plants. Plant J 20:171–182

Bizily SP, Rugh CL, Meagher RB (2000) Phytodetoxification of hazardous organomercurials by genetically engineered plants. Nature Biotechnol 18:213–217

Blaylock MJ (2000) Field demonstrations of phytoremediation of lead contaminated soils. In: Terry N, Bañuelos G (eds) Phytoremediation of contaminated soil and water. Lewis Publishers, Boca Raton, pp 1–12

Brewer EP, Saunders JA, Angle JS, Chaney RL, McIntosh MS (1999) Somatic hybridization between the zinc accumulator *Thlaspi caerulescens* and *Brassica napus*. Theor Appl Genet 99:761–771

Chaney RL, Angle JS, McIntosh MS et al (2005) Using hyperaccumulator plants to phytoextract soil Ni and Cd. Z Naturforsch 60:190–198

Cherian S, Oliveira M (2005) Transgenic plants in phytoremediation: recent advances and new possibilities. Environ Sci Technol 39:9377–9390

Clemens S, Palmgren M, Krämer U (2002) A long way ahead: understanding and engineering plant metal accumulation. Trends Plant Sci 7:309–315

Courbot M, Willems G, Motte P, Arvidsson S, Roosens N, Saumitou-Laprade P, Verbruggen N (2007) A major quantitative trait locus for cadmium tolerance in *Arabidopsis halleri* colocalizes with HMA4, a gene encoding a heavy metal ATPase. Plant Physiol 144: 1052–1065

de Souza MP, Pilon-Smits EAH, Lytle CM et al (1998) Rate-limiting steps in selenium volatilization by *Brassica juncea*. Plant Physiol 117:1487–1494

De Souza MP, Huang CPA, Chee N, Terry N (1999) Rhizosphere bacteria enhance the accumulation of selenium and mercury in wetland plants. Planta 209:259–263

De Souza MP, Lytle CM, Mulholland MM, Otte ML, Terry N (2000) Selenium assimilation and volatilization from dimethylselenoniopropianate by Indian mustard. Plant Physiol 122: 1281–1288

Dhankher OP, Li Y, Rosen BP (2002) Engineering tolerance and hyperaccumulation of arsenic in plants by combining arsenate reductase and γ-glutamylcysteine synthetase expression. Nat Biotechnol. doi:10.1038/nbt747

Doty SL (2008) Enhancing phytoremediation through the use of transgenics and endophytes. New Phytol 179:318–333

Dushenkov S, Kapulnik Y (2000) Phytofiltration of metals. In: Raskin I, Ensley BD (eds) Phytoremediation of toxic metals. Using plants to clean up the environment. Wiley, New York, pp 89–106

Eapen S, D'Souza S (2005) Prospects of genetic engineering of plants for phytoremediation of toxic metals. Biotechnol Adv 23:97–114

Evans K, Gatehouse J, Lindsay W et al (1992) Expression of the pea metallothionein-like gene PsMTa in *Escherichia coli* and *Arabidopsis thaliana* and analysis of trace metal ion accumulation: implications for PsMTa function. Plant Mol Biol 20:1019–1028

Farinati S, DalCorso G, Bona E, Corbella M, Lampis S, Cecconi D, Polati R, Berta G, Vallini G, Furini A (2009) Proteomic analysis of *Arabidopsis halleri* shoots in response to the heavy metals Cadmium and Zinc and rhizosphere microorganisms. Proteomics 9:4837–4850

Farinati S, DalCorso G, Varotto S, Furini A (2010) The *Brassica juncea* BjCdR15, an ortholog of *Arabidopsis* TGA3, is a regulator of cadmium uptake, transport and accumulation in shoots and confers cadmium tolerance in transgenic plants. New Phytol 185:964–978

Farinati S, DalCorso G, Panigati M, Furini A (2011) Interaction between selected bacterial strains and *Arabidopsis halleri* modulates shoot proteome and cadmium and zinc accumulation. J Exp Bot 62:3433–3447

Frey B, Zierold K, Brunner I (2000) Extracellular complexation of Cd in the Hartig net and cytosolic Zn sequestration in the fungal mantle of *Picea abies*—*Hebeloma crustuliniforme* ectomycorrhizas. Plant Cell Environ 23:1257–1265

Fusco, N. Micheletto L, DalCorso G, Borgato L, Furini A (2005) Identification of cadmium-regulated genes by cDNA-AFLP in the heavy metal accumulator *Brassica juncea* L. J Exp Bot 56:3017–3027

Glass DJ (1999) U.S. and international markets for phytoremediation, 1999–2000. D Glass Associates, Needham

Glick BR (2003) Phytoremediation: synergistic use of plants and bacteria to clean up the environment. Biotechnol Adv 21:383–393

Gong J, Lee DA, Schroeder JI (2003) Long-distance root-to-shoot transport of phytochelatins and cadmium in *Arabidopsis*. Proc Natl Acad Sci U S A 100:10118–10123

Hanikenne M, Talke IN, Haydon MJ, Lanz C, Nolte A, Motte P, Kroymann J, Weigel D, Krämer
    U (2008) Evolution of metal hyperaccumulation required *cis*-regulatory changes and
    triplication of HMA4. Nature 453:391–395
Kohler C, Merkle T, Neuhaus G (1999) Characterization of a novel gene family of putative cyclic
    nucleotide- and calmodulin-regulated ion channels in *Arabidopsis thaliana*. Plant J 18:97–104
Korenkov V, Park S, Cheng N et al (2007) Enhanced Cd^{2+} - selective root-tonoplast-transport in
    tobaccos expressing *Arabidopsis* cation exchangers. Planta 225:403–411
Krämer U, Talke I, Hanikenne M (2007) Transition metal transport. FEBS Lett 581:2263–2272
LeDuc DL, Tarun AS, Montes-Bayón M et al (2004) Overexpression of selenocysteine
    methyltransferase in *Arabidopsis* and Indian mustard increases selenium tolerance and
    accumulation. Plant Physiol 135:377–383
LeDuc DL, AbdelSamie M, Montes-Bayón M et al (2006) Overexpressing both ATP sulfurylase
    and selenocysteine methyltransferase enhances selenium phytoremediation traits in Indian
    mustard. Environ Pollut 144:70–76
Li ZS, Lu YP, Zhen RG et al (1997) A new pathway for vacuolar cadmium sequestration in
    Saccharomyces cerevisiae: YCF1-catalyzed transport of bis (glutathionato) cadmium. Proc
    Natl Acad Sci U S A 94:42–47
Martínez M, Bernal P, Almela C et al (2006) An engineered plant that accumulates higher levels
    of heavy metals than *Thlaspi caerulescens,* with yields of 100 times more biomass in mine
    soils. Chemosphere 64:478–485
Meharg AA, Macnair MR (1992) Genetic correlation between arsenate tolerance and the rate of
    influx of arsenate and phosphate in *Holcus lanatus*. Heredity 69:336–341
Moreno FN, Anderson CWN, Stewart RB, Robinson BH (2005) Mercury volatilization and
    phytoextraction from base-metal mine tailings. Environ Pollut 136:341–352
Nedelkoska TV, Doran PM (2000) Hyperaccumulation of cadmium by hairy roots of *Thlaspi
    caerulescens*. Biotechnol Bioeng 67:607–615
Peterson AG, Oliver DJ (2006) Leaf-targeted phytochelatin synthase in *Arabidopsis thaliana*.
    Plant Physiol Biochem 44:885–892
Pilon-Smits E (2005) Phytoremediation. Annu Rev Plant Biol 56:15–39
Pilon-Smits EA, LeDuc DL (2009) Phytoremediation of selenium using transgenic plants. Curr
    Opin Biotechnol 20:207–212
Pilon-Smits E, Hwang SB, Lytle CM et al (1999) Overexpression of ATP sulfurylase in *Brassica
    juncea* leads to increased selenate uptake, reduction and tolerance. Plant Physiol 119:123–132
Pomponi M, Censi V, Di Girolamo V, De Paolis A, Sanità di Toppi L, Aromolo R, Costantino P,
    Cardarelli M (2006) Overexpression of *Arabidopsis* phytochelatin synthase in tobacco plants
    enhances Cd^{2+} tolerance and accumulation but not translocation to the shoot. Planta
    223:180–190
Rogers EE, Eide DJ, Guerinot ML (2000) Altered selectivity in an *Arabidopsis* metal transporter.
    Proc Natl Acad Sci U S A 97:12356–12360
Ruiz ON, Alvarez D, Torres C, Roman L, Daniell H (2011) Metallothionein expression in
    chloroplasts enhances mercury accumulation and phytoremediation capability. Plant
    Biotechnol J 9:609–617
Rugh CL, Wilde HD, Stack NM, Thompson DM, Summers AO, Meagher RB (1996) Mercuric
    ion reduction and resistance in transgenic *Arabidopsis thaliana* plants expressing a modified
    bacterial merA gene. Proc Natl Acad Sci U S A 93:3182–3187
Salt DE, Smith RD, Raskin I (1998) Phytoremediation. Ann Rev Plant Physio 49:643–668
Sasaki Y, Hayakawa T, Inoue C et al (2006) Generation of mercuryhyperaccumulating plants
    through transgenic expression of the bacterial mercury membrane transport protein MerC.
    Transgenic Res 15:615–625
Sauge-Merle S, Cuiné S, Carrier P, Lecomte-Pradines C, Luu DT, Peltier G (2003) Enhanced
    toxic metal accumulation in engineered bacterial cells expressing *Arabidopsis thaliana*
    phytochelatin synthase. Appl Environ Microbiol 69:490–494

Schuurink RC, Shartzer SF, Fath A, Jones RL (1998) Characterization of a calmodulin-binding transporter from the plasma membrane of barley aleurone. Proc Natl Acad Sci U S A 95:1944–1949

Shaw WH, Anderson JW (1972) Purification, properties and substrate specificity of adenosine triphosphate sulphurylase from spinach leaf tissue. Biochem J 127:237–247

Song W, Sohn EJ, Martinoia E, Lee YJ et al (2003) Engineering tolerance and accumulation of lead and cadmium in transgenic plants. Nat Biotechnol 21:914–919

Talke I, Hanikenne M, Krämer U (2006) Zinc dependent global transcriptional control, transcriptional de-regulation and higher gene copy number for genes in metal homeostasis of the hyperaccumulator *Arabidopsis halleri*. Plant Physiol 142:148–167

Thomas JC, Davies EC, Malick FK, Endreszl et al (2003) Yeast metallothionein in transgenic tobacco promotes copper uptake from contaminated soils. Biotechnol Prog 19:273–280

van der Zaal BJ, Neuteboom LW, Pinas JE et al (1999) Overexpression of a novel *Arabidopsis* gene related to putative zinc-transporter genes from animals can lead to enhanced zinc resistance and accumulation. Plant Physiol 119:1047–1055

van Huysen T, Abdel-Ghany S, Hale KL et al (2003) Overexpression of cystathionine-g-synthase in Indian mustard enhances selenium volatilization. Planta 218:71–78

Vatamaniuk OK, Mari S, Lu Y, Rea PA (2000) Mechanism of heavy metal ion activation of phytochelatin (PC) synthase. Biol Chem 275:31451–31459

Verbruggen N, Hermans C, Schat H (2009) Molecular mechanisms of metal hyperaccumulation in plants. New Phytol 181:759–776

Wernick I, Themelis N (1998) Recycling metals for the environment. Annu Rev Energy Environ 23:465–497

Wijnhoven S, Leuven R, Van Der Velde G et al (2007) Heavy-metal concentrations in small mammals from a diffusely polluted floodplain: importance of species- and location-specific characteristics. Arch Environ Contam Toxicol 52:603–613

Wojas S, Clemens S, Hennig J, Skłodowska A, Kopera E, Schat H et al (2008) Overexpression of phytochelatin synthase in tobacco: distinctive effects of AtPCS1 and CePCS genes on plant response to cadmium. J Exp Bot 59:2205–2219

Zayed A, Lytle CM, Terry N (1998) Accumulation and volatilization of different chemical species of selenium by plants. Planta 206:284–292

Zhu YL, Pilon-Smits EA, Jouanin L, Terry N (1999a) Overexpression of glutathione synthetase in Indian mustard enhances cadmium accumulation and tolerance. Plant Physiol 119:73–80

Zhu YL, Pilon-Smits EHA, Tarum AS et al (1999b) Cadmium tolerance and accumulation in Indian mustard is mnhanced by overexpressing γ-Glutamylcysteine synthetase. Plant Physiol 121:1169–1177